UNDERSTANDING GEOLOGY

David Webster

on or before

Oliver & Boyd

Acknowledgements

The publishers thank the following for permission to reproduce photographs:

NASA (2.2, 2.5, 2.8)
Popperfoto (2.11)
NASA/Science Photo Library (2.15)
British Museum (Natural History) (2.19, 2.20, 3.1, 3.10, 3.11, 5.13, 5.17, 7.10, 7.15, 7.16, 7.21a, b, 8.10, 9.3, 12.12, 12.18, 12.44, 15.3)
Geoscience Features (5.5, 5.7, 5.8, 12.30, 12.40)
David Webster (3.5a, 5.6, 6.3, 6.5, 6.39, 7.3, 7.4, 7.23, 7.30, 7.33, 9.5, 10.2, 10.3, 10.10, 10.16, 10.21, 10.27, 11.2, 12.2, 12.22, 14.1, 14.16, 14.18, 14.23, 14.24, 15.16)
RTZ (3.5b)
Aerofilms (5.1, 6.19, 6.31, 6.33, 6.36, 6.40, 7.11, 9.7, 14.26, 15.7)
Landform Slides (5.9, 6.2, 6.23, 6.44, 14.29, 14.30)
Reproduced by permission of the Director, British Geological Survey (NERC):
Crown/NERC copyright reserved (6.4, 7.25, 7.26, 7.31, 12.32, 14.8)
Eric Kay (6.9, 6.12a, 14.7)
Tony Waltham (6.12b, 6.24, 13.3, 14.11, 14.15)
Rochdale WEA Geology class (6.26)
J M Hancock (7.14)
Rida (7.18 David Bayliss, 12.1 Richard Moody)
Mary Evans Picture Library (12.6)
Seaphot (12.23, 12.34, 12.37)
US Geological Survey (13.13)
Harry M Parker (15.4)
International Tin Research Institute (15.5)
British Coal (15.9)
Shell (15.13)

*Illustrated by Tim Smith, David Hogg, Alan Timms (invertebrate fossil diagrams),
Ann Rooke and Cauldron Design Studio*

The author thanks all friends and colleagues who have assisted in the preparation of this book, in particular Fred Broadhurst, Alan Timms and Peter Kennett for their most helpful suggestions on aspects of the text. He also thanks Jenny Kilvert who loaned some of the specimens used in the photographs. Finally, a special mention must be given to Nina, Matthew, Helena, Peter and Alexa to whom this book is dedicated.

Oliver & Boyd
Edinburgh Gate
Harlow
Essex CM20 2JE

An Imprint of Longman Group UK Ltd

ISBN 0 05 003664 5
First Published 1987
Ninth impression 1995

Set in 11/12pt Univers 45 and 65

Produced by Longman Singapore Publishers Pte Ltd
Printed in Singapore
WLEE/08

The publisher's policy is to use paper manufactured
from sustainable forests.

Contents

1 What is geology?

Let us begin with some dictionary definitions:

'geo' comes from a Greek word meaning *earth*,
geology means the *study of the earth*,
geologist means a *person who studies the earth*.

Geology can also be called **earth science**. Since it is concerned with a whole planet, it is a very broad subject which involves such questions as:

How was the earth formed?
What is the earth made of?
How has the earth changed since its formation?
How has life on earth evolved?
Where can important metals and fuels be found in the earth?
How is the earth affected by people using its natural resources?

Many of these questions are difficult to answer because they are about things that happened millions of years ago or at a very slow rate. To answer them, geologists need to be scientists and detectives. They must study the clues left by events from the past and compare them with things that can be seen happening at the present time. They must also test their theories by setting up suitable experiments. For example, geologists can only explain how sandstones formed in ancient deserts by studying modern dunes and carrying out experiments to see what wind speeds are needed to move different sizes of sand grain.

Only by linking together many clues and separate pieces of information can we begin to explain what has occurred during the vast history of the earth.

There is perhaps one weakness with the term 'geology' and its meaning of 'earth study'. When the word was first used in the eighteenth century, nobody imagined that information from other planets would ever be available. Today, with the development of space exploration, geologists are beginning to study specimens and photographs from beyond the earth. Perhaps it is time to rename our subject **planetology**?

Fig. 1.1 Becoming a geologist: a geologist studies the earth in a logical way

2 Planet Earth

Before studying the earth itself let us look beyond our planet to its neighbours in space.

The Solar System

The earth is one of nine planets which circle around the sun (see Fig. 2.1). The sun is our 'local' star and, without its heat and light, there would be no life on earth. The sun has a diameter of about 1 500 000 km (100 times that of the earth), weighs about 330 000 times as much as the earth and produces vast amounts of energy. We can feel its heat even though we are 150 million km away from it. The sun and its nine planets make up the **Solar System** ('sol' is the Latin word for sun).

Mercury, Venus, Earth and Mars are the **inner planets** which are generally small in size and made of solid rock. Jupiter, Saturn, Uranus, Neptune and Pluto are the **outer planets**. They tend to be larger, less dense and made of liquids and gases. Between Mars and Jupiter there is an **asteroid belt**. Asteroids are rocky particles which failed to develop into a planet when the Solar System was formed.

Mercury Extremely hot with no atmosphere. Space probes have discovered mountains and craters on its surface.
Venus Surface hidden by swirling clouds but spacecraft pictures show a rocky landscape. Surface temperature about 480 °C.
Mars Orange in colour. Photographs show huge volcanoes and valleys. Atmosphere believed to be of

Fig. 2.1 The Solar System: showing the distance of each planet from the sun, and the time each planet takes to orbit the sun

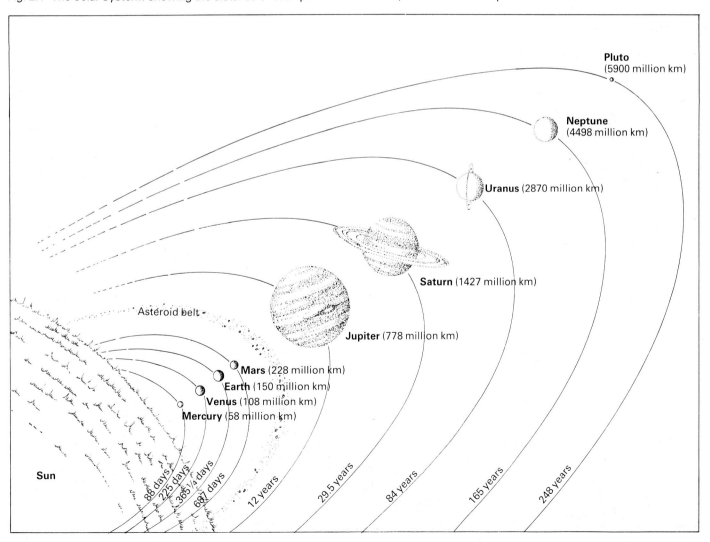

carbon dioxide gas with the polar areas covered by frozen carbon dioxide and ice. Surface temperature about −23 °C.

Jupiter The largest planet in the system. Its atmosphere appears banded and contains a massive red spot believed to be a constantly whirling storm. Surface temperature −150 °C. Surrounded by 13 'moons'.

Saturn Cloudy atmosphere. Orbited by four rings of ice-covered particles and 10 'moons'. The rings could represent a 'moon' which has broken up.

Uranus Rotates in such a way that some parts are in darkness for periods of up to 21 years. Has a green atmosphere probably of methane.

Neptune and **Pluto** So far away that information is very difficult to collect. Surface temperatures are probably colder than −200 °C. Pluto may not even be a proper planet but a 'moon' which has escaped from orbiting somewhere else in the Solar System.

In Fig. 2.1 the distances between the planets are not drawn to scale. The planets should be much farther apart. To get an idea of the true scale of the Solar System you would have to

- put a football on the ground to represent the sun,
- place a pea 33 metres from it (Earth),
- place an orange 175 metres from it (Jupiter),
- place a pin head 1.5 km from it (Pluto).

The Solar System is only one of millions of other star systems in our **galaxy** which is called the **Milky Way**. In turn there are countless other galaxies within the whole **universe**. Many questions remain unanswered about the universe, and in some ways the ideas involved are beyond human understanding. One obvious question is 'does the universe end?'. Can you believe that space ends somewhere? On the other hand can you believe that it goes on for ever?

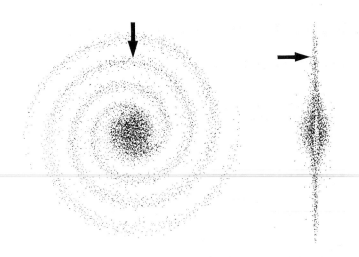

Fig. 2.3 Two views to show what the huge Milky Way galaxy might look like. It is 1 million million million kilometres across. The arrows show the position of the Solar System within the Milky Way

Units of measurement such as the mile or kilometre are too small to use in the vastness of space so distances are calculated in **light years**. A light year is the distance travelled by a beam of light in one year. Since light travels at 300 000 km per second, a light year is about 10 million million km. Light takes just over 8 minutes to travel from the sun to earth. The next nearest star is 4 light years from earth, while the Milky Way is 100 000 light years across.

When you look at some of the most distant stars you are seeing light which began travelling thousands or millions of years in the past. In a sense you are looking back in time.

Whether humans will ever travel beyond our Solar System depends on the speed of future spacecraft. For journeys within the Milky Way we would need to travel faster than light, otherwise we would not get to our destination (and back?) within a lifetime.

Fig. 2.2 Olympus Mons: a volcano on Mars; 25 km high and 600 km across, the summit can be seen rising above the clouds

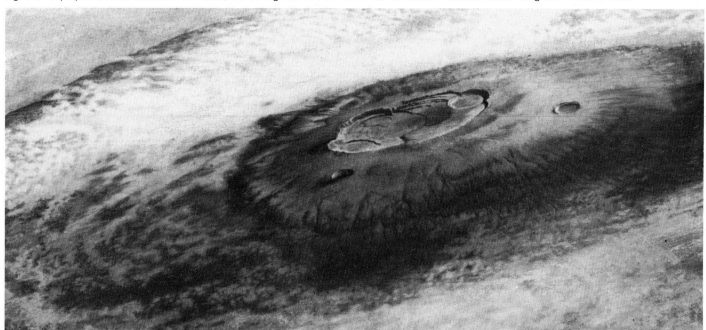

Planet Earth: facts and figures

The shape of the earth

The earth is not an exact sphere. Its spinning motion produces a slight flattening at the poles and a slight bulging at the equator. Compare these two figures:

diameter at the equator = 12 757 km,
diameter at the poles   = 12 714 km.

What is the difference between these figures?
What is the average diameter of the earth?
What is the earth's circumference?

The density of the earth

Mass of the earth
  = 5 980 000 000 000 000 000 000 000 000 grams (g).

Volume of the earth
  = 1 083 000 000 000 000 000 000 000 000 cubic centimetres (cm^3)

Density of earth = $\dfrac{mass}{volume}$ = 5.5 g/cm^3.

This calculation gives the average density of the whole earth. Most common rocks found at the earth's surface have a much lower density of about 2.8 g/cm^3.
What does this tell you about the density of the rocks deep within the planet?

The surface of the earth

Land area = 148 million km^2
(29% of surface)

Sea area = 362 million km^2
(71% of surface)

Largest continent = 54 million km^2 (Eurasia)

Largest ocean = 179 million km^2 (Pacific)

Average height of land = 850 m above sea level

Average depth of oceans = 3500 m below sea level

Highest mountain peak = +8848 m (Mount Everest, Himalayas)

Deepest ocean floor = −10914 m (Marianas Trench, Pacific)

Fig. 2.5 Earth from space, described by astronauts as a 'living green' colour. *What continent can you recognise?*

Fig. 2.4 shows that, when oceans, forests, mountains, ice sheets and deserts are taken into account, only a small percentage of the earth's surface is left to support the world's population.

The age of the earth

Calculations have been made to find the age of the earth, the age of moon rocks collected by astronauts and the age of meteorites which fall to earth from space. These calculations depend on accurately measuring the amount of decay that has taken place in certain radioactive elements since they were first formed (the technique is explained further on page 90). Evidence suggests that the whole Solar System first came into being about **4600 million years ago**.

By comparison, evidence for the first life on earth (bacteria) comes from rocks over 3000 million years old. The dinosaurs 'ruled the earth' about 130 million years ago, and the first real humans appeared less than 2 million years ago.

Fig. 2.4 Bar graph of the earth's surface

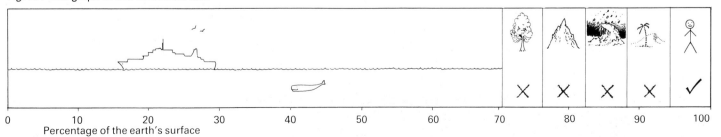

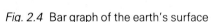
Percentage of the earth's surface

Is the earth unique?

The earth supports life, and one vital fact in the development of this life is the distance from the earth to the sun. This distance is such that the heat from the sun allows water on the earth's surface to exist as a liquid.

What would happen to the earth's water and the life it supports if the earth was (a) much closer to the sun, (b) much further away from the sun?

Another important feature of the earth is its size. It is large enough to 'hold down' an atmosphere by gravity and also to produce heat energy (by natural radioactive decay) within its interior. The atmosphere provides life-supporting gases and shields us from solar radiation. The internal heat provides energy for much of the geological activity (for example, volcanoes, mountain building, etc.) which constantly affects the planet.

The activity of the earth is in sharp contrast to that of its nearest neighbour in space, the **moon** (Fig. 2.7), which is too small to maintain an atmosphere or generate much internal energy. The moon is a 'dead planet' with hardly any geological activity and where no life forms can exist, despite the fact that it is about the same distance from the sun as the earth is.

Thinking about other planets which may exist beyond the Solar System presents some very interesting questions.

Could there be other earth-like planets? Are there other life forms? Is the earth unique? What do you think?

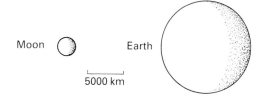

Fig. 2.7 The relative size of the earth and the moon

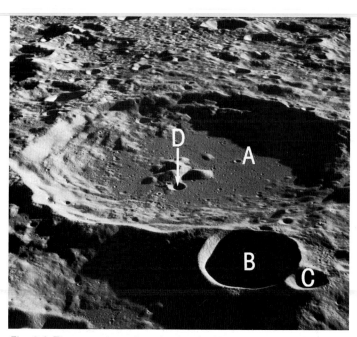

Fig. 2.8 The moon's surface is pitted with many craters caused by the impacts of meteorites. *Which of the four craters (A, B, C or D) do you think is the oldest? Give reasons for your answer.*

Fig. 2.6 The development of the earth

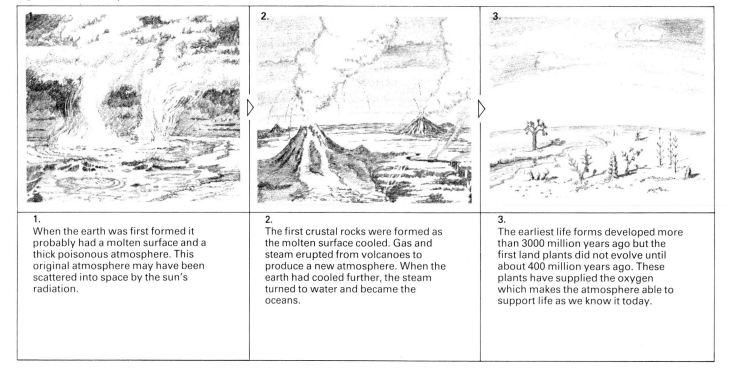

1.
When the earth was first formed it probably had a molten surface and a thick poisonous atmosphere. This original atmosphere may have been scattered into space by the sun's radiation.

2.
The first crustal rocks were formed as the molten surface cooled. Gas and steam erupted from volcanoes to produce a new atmosphere. When the earth had cooled further, the steam turned to water and became the oceans.

3.
The earliest life forms developed more than 3000 million years ago but the first land plants did not evolve until about 400 million years ago. These plants have supplied the oxygen which makes the atmosphere able to support life as we know it today.

Inside the earth

Detailed information about the earth's interior cannot come from mining or drilling operations. The world's deepest mine, in the South African goldfields, reaches a depth of nearly 3.5 km, but this is only a surface scratch compared with the planet's radius of over 6000 km. The mine does, however, prove that temperature increases with depth, because at the deepest point miners have to endure temperatures of over 50 °C.

Most evidence about the really deep interior comes from the study of **earthquakes**. Further information is gathered from **volcanoes, meteorites** and by making mathematical calculations.

Fig. 2.9 The zones of the earth

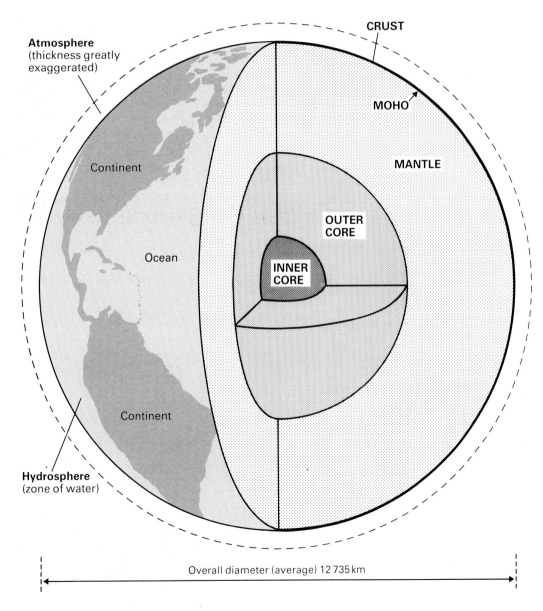

Atmosphere (thickness greatly exaggerated)

Continent

Ocean

Continent

Hydrosphere (zone of water)

CRUST

MOHO

MANTLE

OUTER CORE

INNER CORE

Overall diameter (average) 12 735 km

Crust: solid rock but very thin compared with the size of the whole earth. Density of 2.7 to 3.2 g/cm^3. May be divided into **continental** and **oceanic crust**. See page 13 for full details.

Boundary between crust and mantle called the Mohorovičić discontinuity or **moho** for short

Mantle: dark-coloured solid rocks rich in magnesium and silicon. Density of 3.4 to 5.5 g/cm^3. Parts of the mantle are slowly moving in a semi-plastic way because of currents produced by the heat of radioactive decay.

Outer core: made of nickel and iron (NiFe), believed to be in a liquid state due to the extremely high temperatures. Density of 10 to 12 g/cm^3. The earth's magnetism is produced in this zone. Diameter of outer core = 6930 km

Inner core: also made of NiFe but in a solid state due to pressures over 3 million times greater than those at the surface. Density between 12 and 18 g/cm^3. Temperatures of over 3000°C? Diameter of inner core = 2530 km

Earthquakes and the earth's interior

The study of earthquakes is called **seismology** (from the Greek word seismos meaning 'to shake'). Earthquakes can cause spectacular damage and loss of life. The greatest death toll occurred in Shanxi Province, China, in 1556 when an estimated 830 000 people were killed. Death and destruction can also result from huge ocean waves known as **tsunami** (pronounced 'soo-narmee'), which are set off by earthquakes. It is estimated that about 500 000 earthquakes occur each year but of these only about 1000 are strong enough to cause any damage and only a few could be classed as serious.

An earthquake occurs when forces inside the earth become strong enough to fracture large masses of rock and make them move. This sudden break releases energy which travels through the earth as a series of **shock waves**. Think what happens when you bend a ruler. If you keep on applying force there comes a point when the ruler can bend no more and so it snaps. When the break occurs the energy you have been applying is suddenly released causing the broken ends to move rapidly apart and shock waves to be given out. You hear some of these shock waves as the snapping sound.

Most earthquakes are generated within 600 km of the earth's surface. It seems that at greater depths the rocks are so hot that they behave in a 'plastic' way. Trying to produce shock waves deep inside the earth would be like trying to make Plasticine break with a sudden snap!

Recording and measuring earthquakes

The exact point where an earthquake originates is called the **focus** (plural: foci), and the nearest point on the earth's surface directly above it is the **epicentre**. Shock waves will be felt most strongly at the epicentre but will also spread out from this point. Some will actually travel right through the inside of the planet.

Earthquake shocks are recorded on a **seismograph**. Fig. 2.10 illustrates a very simple type of machine. Most modern seismographs use more complicated methods, making them sensitive enough to detect faint shock waves thousands of kilometres from the epicentre. However, all seismographs rely on the same principle: one part of the instrument shakes with the earth while another part remains still.

Reports from areas affected by earthquakes often give figures to describe the size of the shock. Two scales are commonly used for this.

1. The Richter scale. This measures energy released, but it is important to realise that each scale point means an energy increase of ten times. For example, a scale 6 earthquake is ten times as powerful as a scale 5, a hundred times as powerful as a scale 4, a thousand times as powerful as a scale 3.

2. The Mercalli scale. This uses Roman numerals from I to XII (1 to 12) to measure strength in terms of damage caused and effects felt. For example, a scale IV (4) earthquake will only cause small objects to rock slightly while a scale XI (11) will leave few buildings standing.

Fig. 2.11 Rescue workers search for survivors after the Mexico City earthquake of September 1985

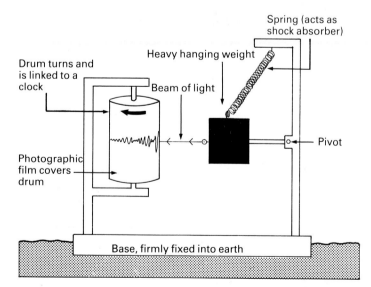

Fig. 2.10 How a simple seismograph works. The machine is designed so that, even when the earth shakes, the heavy weight (and its light) tend to remain still. As the drum shakes with the earth, it moves up and down against the steady beam of light and so records the pattern of shock waves onto the film

Fig. 2.12 A seismograph record

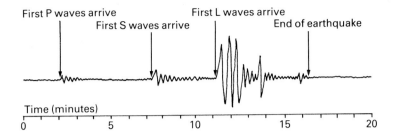

First P waves arrive
First S waves arrive
First L waves arrive
End of earthquake

Time (minutes)
0   5   10   15   20

Types of earthquake waves

Fig. 2.12 shows a typical seismograph record. Three types of shock waves are recorded. Each type has its own special features, as described below. Use the key letters to help you remember them.

P waves These **p**ush waves travel within the earth and **p**ass through both solids and liquids. They travel faster than other types and are therefore the **p**rimary ones to be recorded.

S waves These **s**hake waves also travel in the interior but are **s**topped by liquids. Being **s**lower than P waves they arrive in **s**econd place.

L waves These **l**ong waves only travel in the crust. Because of their **l**arge movement they produce the most damage. Being the slowest wave type they are always **l**ast to be recorded.

Seismographs and the earth's interior

Using earthquake waves and seismographs to investigate the interior of the earth is rather like tapping something to find out what is inside. For example, if you tap on a barrel you are sending mild shock waves through it. These waves will behave differently, and therefore make a different sound, depending on whether the barrel is empty, full of liquid or full of something solid. Similarly the pattern of waves received by a seismograph depends on what they have passed through inside the earth.

Fig. 2.14 shows what happens when shock waves meet a boundary between two different materials. (It is similar to what happens to light waves when they pass from air into water.) Notice that the waves are **reflected, refracted** and their velocity (speed) is changed.

The pattern of waves received by a seismograph is usually more complex than Fig. 2.12 shows. This is because P and S waves have been reflected, refracted and changed in velocity as they have travelled within the earth. For waves to be affected in this way they must have passed through a series of boundaries separating zones of different materials. Fig. 2.9 shows the internal structure of the earth. The position

Fig. 2.13

WONDER WHAT'S IN THIS

Fig. 2.14 Reflection and refraction of shock waves

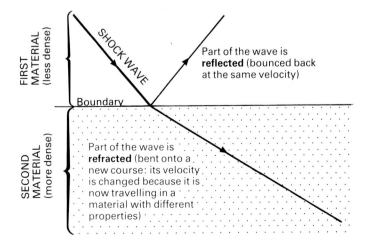

FIRST MATERIAL (less dense)

SHOCK WAVE

Part of the wave is **reflected** (bounced back at the same velocity)

Boundary

SECOND MATERIAL (more dense)

Part of the wave is **refracted** (bent onto a new course: its velocity is changed because it is now travelling in a material with different properties)

of boundaries (such as the **moho**) and the thicknesses and densities of the zones have all been worked out from the way earthquake shock waves have been affected by passing through them.

S waves are particularly affected by the zone known as the **outer core**. In fact they do not pass through it at all. Since it is known that S waves will not travel in liquids we can assume that the outer core is in a liquid state (probably due to the great heat at that depth).

It is not only the deep interior of the planet that can be studied using seismic methods. Geologists searching for oil or other valuable deposits in the earth's crust can produce 'artificial' shock waves by detonating small explosive charges. They then record the wave patterns on small seismographs and process the information on computers. In this way the position, properties and thickness of rock layers can be found without the expense of drilling boreholes.

Other information about the earth's interior

Apart from earthquake data, other sources give us important information about the inside of the earth.

Volcanoes. The **lava** which is erupted by volcanoes comes from inside the earth, but this lava is usually produced within the crust or upper mantle so it does not tell us much about the really deep layers within the earth. The distribution of volcanoes shows the areas of the earth's interior where heat is being generated and rising. You can learn more about volcanoes on pages 31–7.

Meteorites. These are particles of dust or larger pieces of rock which have been drawn towards the earth from space (by the earth's gravity). Most meteorites burn up as they pass through the atmosphere but occasionally they fall to the earth's surface. Since they may be samples of planetary material left over from the time of formation of the Solar System they are of interest to geologists. One type of meteorite, for instance, is rich in iron and nickel, and may be of similar material to the earth's core.

The earth's magnetic field. The fact that the planet possesses a magnetic field can be taken as further evidence that the earth's core is rich in iron.

Calculations. These can be made to estimate conditions within the earth. For example, if we know temperatures rise by 20 °C during the descent of a 2.5 km deep mine, we can estimate what temperature could be expected if we descended 3000 km to the edge of the core. In the same way it should be possible to calculate the pressure at a particular depth by working out the weight of rocks above. Unfortunately, calculations of this type may prove inaccurate since temperature and pressure are unlikely to increase evenly with depth throughout the whole earth.

Project Mohole

The purpose of this American project was to drill through the earth's crust and, for the first time, study the moho and upper mantle by direct sampling. A borehole site was chosen on the sea floor, and much research was carried out. However, the engineering difficulties and high costs were beyond even the U.S.A., so the project was abandoned in 1967. Although we have since had moon rock specimens from 400 000 km out in space, we have yet to sample as far as even 4 km down into our own planet.

Why do you think an ocean floor site was chosen for project Mohole? (Fig. 2.16 may help you).
What do you think might have happened if the mantle had been 'punctured' by a drill bit?

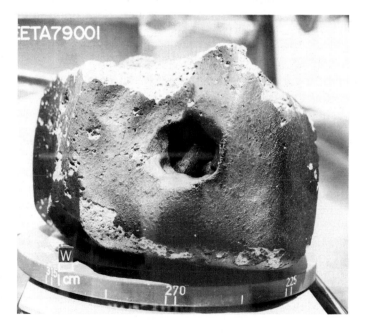

Fig. 2.15 A meteorite discovered in 1981, possibly of Martian origin

The earth's crust

The **crust** is the outer skin of the earth. Its thickness can be compared to that of a postage stamp stuck onto the outside of a football. Geologists know more about the crust than other zones of earth because they can collect rock samples from it. The crust can be divided into two types.

Continental crust

This type of crust forms the earth's continents, and continues a little beyond their edges as **continental shelves** and **continental slopes** under the sea. The upper part is often called **granitic crust** because the average composition of the various rocks here is the same as the rock type **granite**. Granitic crust has been formed over a long time. Rocks dating back to 3700 million years have been found. The deeper parts of the continents are made of a denser material which is thought to be similar to oceanic crust. The total thickness of continental crust varies greatly and can be as much as 70 km beneath high mountain ranges.

Oceanic crust

This forms a much thinner (average 6 km) layer of crust beneath deep sea floors. Since its composition is equal to that of the rock type **basalt**, this material is often called **basaltic crust**. It is relatively young. Nowhere have rocks older than 220 million years been discovered. As mentioned above, similar material to oceanic crust is thought to lie beneath the continents. However, this zone is thicker (and possibly older) than the true basaltic ocean floor.

Mohorovičić discontinuity

The boundary between the crust and mantle is called the **Mohorovičić** (pronounced 'mohoroveechy') **discontinuity**, or **moho** for short. Fig. 2.16 shows that, since the thickness of the crust varies, the depth of the moho must also vary.

The average density of granitic crust is 2.7 g/cm^3, while the density of basaltic crust is 3.0 g/cm^3.

Fig. 2.16 Section through part of the earth's crust and upper mantle. (Note that the vertical scale is very much greater than the horizontal.)

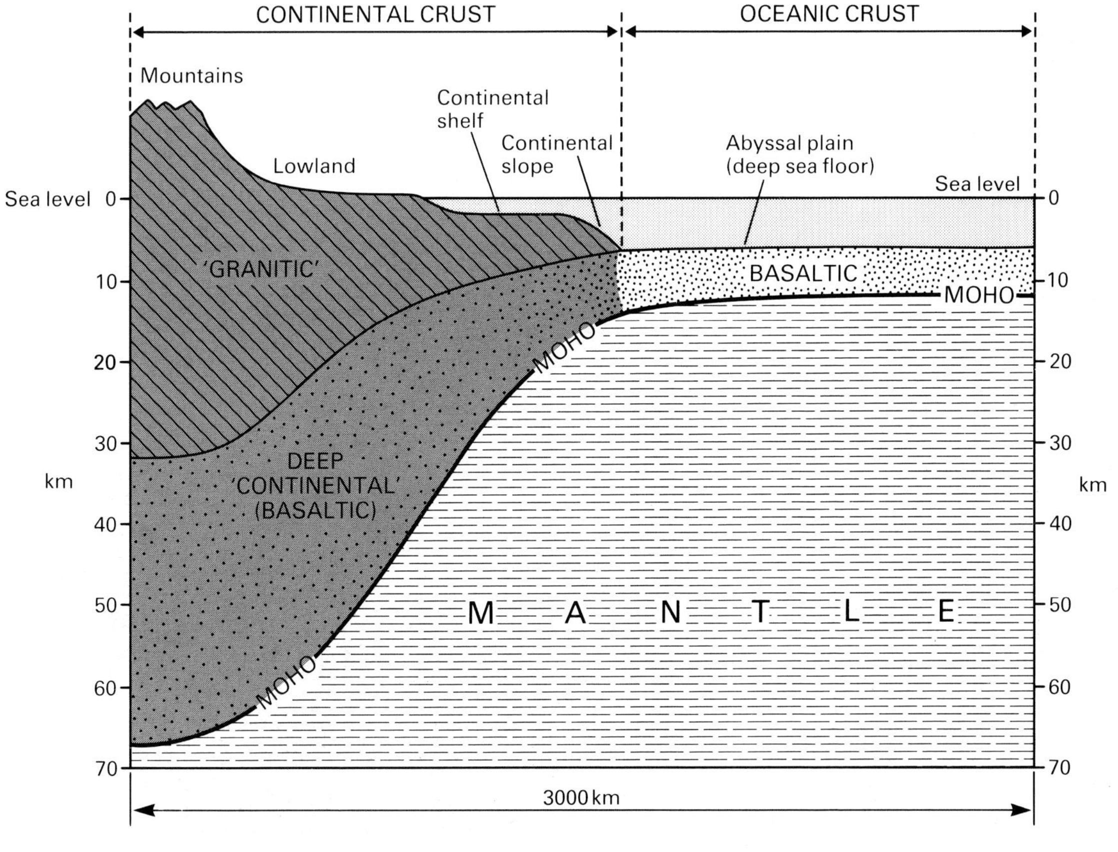

13

Activity in the earth's crust

For over a hundred years geologists have known that **earthquakes** and **volcanoes** tend to occur in certain parts of the earth's crust. Fig. 2.17 shows the 'zones of activity' which exist today. The reasons behind this pattern took a long time to be discovered. In the past, vague reasons such as 'where the crust is thin' or 'where the crust is weak' were given. However, since the 1960s, new research has produced a scientific and logical explanation. Before considering this, let us use Fig. 2.17 and an atlas to study the pattern of activity within the crust.

(a) Earthquakes and volcanoes are mainly confined to 'belts' of activity which stretch across the globe. Most belts are a few hundred kilometres in width but thousands of kilometres in length.

(b) Places can be found where these belts appear to join together (for example, in the middle of the Indian Ocean).

(c) On land the belts occur along chains of high mountains (for example, the Andes, Alps, Himalayas, etc.).

(d) Beneath the seas the belts are either to be found along the centres of oceans (for example, in mid-Atlantic) or they pass through chains of volcanic islands (for example, in the Caribbean and Philippines areas).

(e) Surveys of the sea floor show that the mid-ocean belts are actually along chains of huge mountains beneath the sea. These mountains are much larger than any mountains on land. For example, Iceland is just the top of one part of the **mid-Atlantic Ridge**; here it rises over 4 km from the ocean floor.

(f) Surveys also show that belts which pass through mountain ranges near the edges of continents, and belts which pass through island chains, are always close to deep trenches in the ocean floor (for example, the Peru–Chile Trench runs parallel to the Andes, and the Marianas Trench runs parallel to some of the Philippine Islands).

Fig. 2.17 The earth's 'active belts'

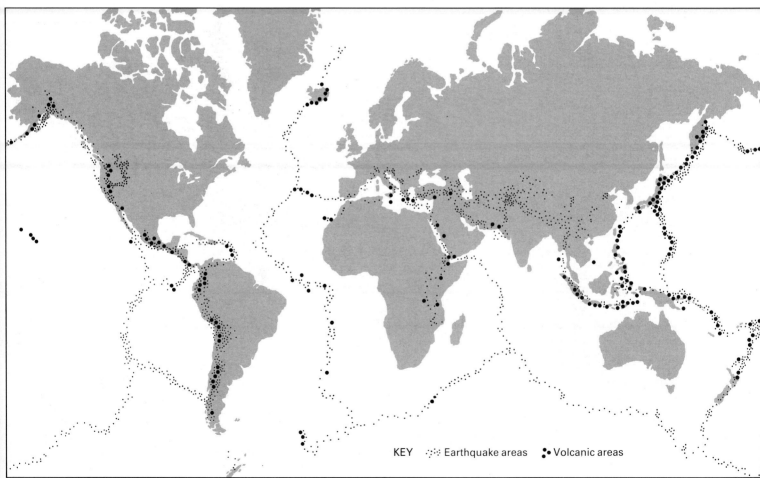

KEY   Earthquake areas   Volcanic areas

An introduction to plate movement

The outer layer of the earth is not quite the continuous solid skin which used to be imagined. In fact it is made up of separate pieces called **plates**. Each plate is a solid 'slab' of **lithosphere** (crust plus the uppermost part of the mantle) some 80–120 km thick. If you compare Figs. 2.17 and 2.18 you will see that the earth's 'active belts' are situated along the edges, or **margins**, of these massive plates.

Each plate moves about the earth's surface at the rate of a few centimetres per year. The first evidence of this movement came in the 1960s when the volcanoes along **oceanic ridges** were studied (oceanic ridges are mountain ranges beneath the sea, for example, at the plate margin in the centre of the Atlantic: see Fig. 2.19 on page 33). It was found that **basalt lava** erupts from these ridges and cools to form new oceanic crust along the edges of the plates on either side (see Fig. 2.21). As this happens, the plates move away from the ridge and the width of the ocean floor is increased. This process is known as **sea-floor spreading**. *Use the arrows on Fig. 2.18 to decide which plate margins are the sites of sea-floor spreading.*

Other studies of the sea floor proved that **oceanic trenches** are where plates meet as they move together (see Fig. 2.20 on page 33). At these places the edge of one plate is forced to slide beneath the other and moves down into the mantle (see Fig. 2.21). A process called **subduction** takes place as the descending plate edge is destroyed or 'digested' by melting at depth. Since subduction involves the destruction of oceanic crust, it follows that some oceans must be shrinking in size. *Use Fig. 2.18 again to find sites where oceanic crust is being subducted.*

The overall effect of the movement described above is to produce a system where each plate acts like a giant 'conveyor belt'. As Fig. 2.21 shows, the 'conveyor' moves slowly from the plate edge where material is being added (an oceanic ridge) to the opposite edge where material is being destroyed (an oceanic trench and subduction zone). Since movement is only a few centimetres per year the 'journey' may take several hundred million years to complete.

In total as much material must be taken away at subduction zones as is added at oceanic ridges. If this were not the case the whole planet would be constantly changing size!

Fig. 2.18 Simplified map of the earth's plates

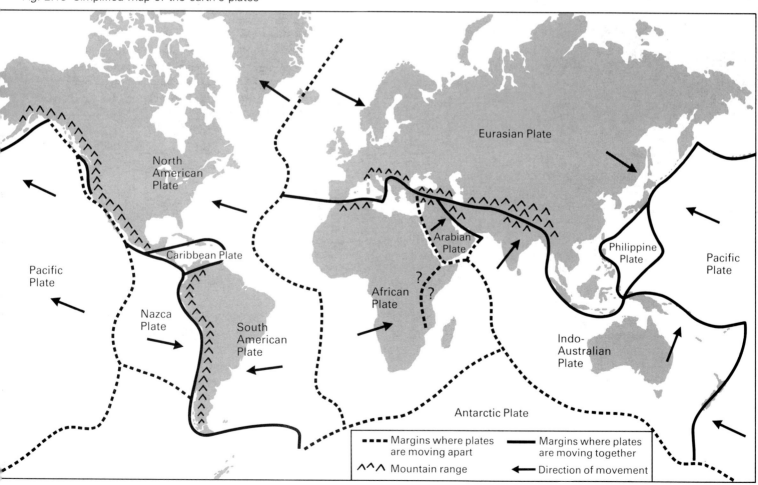

Fig. 2.21 The 'conveyor belt' idea of plate movement. *Note:* the lithosphere (or plate) is made of upper mantle (shaded with horizontal dashes on the diagram) plus the oceanic or contintental crust. If subduction continues, continent A will eventually collide with continent B to produce the situation shown in (b)

(a)

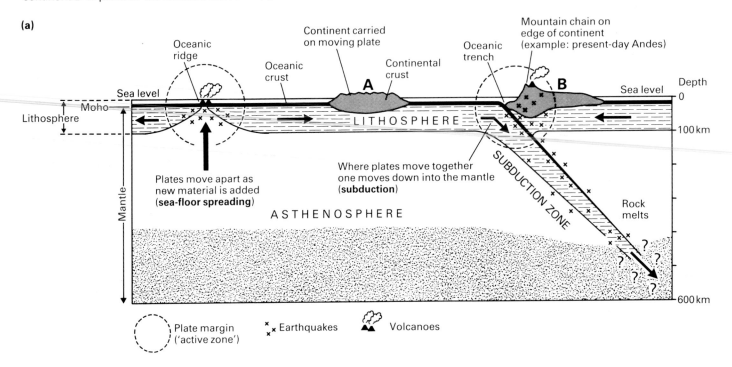

Plates, continents and mountains

As Fig. 2.18 shows, most plates have continents on them. A continent is really a mass of lighter granitic crust resting on the denser basaltic plate material below. In effect, continents 'ride the conveyor belt' beneath them (see Fig. 2.21). As the plates move, the continents which they carry also move. Although the movement may seem too slow to have much effect, do not forget that it has happened over thousands of millions of years. The continents have already travelled many thousands of kilometres around the globe. Northern Europe, for example, was near the equator about 300 million years ago. Tropical forests grew there then, which later formed our coal deposits.

The movement of continents is also responsible for the formation of mountain ranges (for example, the Andes and Himalayas). This happens when plate movement carries continents towards subduction zones. Granitic crust is not dense enough to descend into the mantle with the rest of the plate so the continental edge nearest the subduction zone is forced up and distorted to form a mountain range.

Look at the illustration of mountain building and subduction in Fig. 2.21 then use Fig. 2.18 to work out the relationship between plate movement and the formation of mountains along the west coast of South America.

(b) **If subduction continues, continent A will eventually collide with continent B to produce the situation shown here**

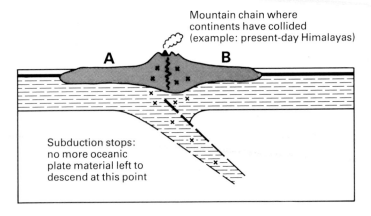

It has already been mentioned that, because of subduction, some oceans are shrinking in size. If this process continues for long enough, a complete ocean floor may be destroyed and two continents may come together and join into one. As their edges collide, the crust is forced up to form a mountain chain. *Use Fig. 2.18 again to see how the Himalayan mountains in India were formed.*

How are the plates on the surface of the earth able to move over the deeper parts of the mantle? It seems to be possible because of a layer known as the **asthenosphere** which lies immediately below the lithosphere plates. Earthquake waves are slowed down by material in the asthenosphere, which suggests that this layer is semi-plastic and 'weaker' than the rest of the mantle. However, the relationship between surface plate movement and movements of the underlying mantle is, as yet, not fully understood.

The subject of plate movement (known as **plate tectonics**) is a vital part of modern geological science. It explains much more than why mountains, trenches, earthquakes and volcanoes occur where they do. It also leads into fascinating topics such as moving continents, changing environments, etc. However, before you can study these topics in more detail, you need to understand the 'basics of geology' (for example, how different rocks are formed, what effects earth forces have, etc.). We shall return to plate tectonics later in the book (page 139), but in the meantime try to keep it in mind as the mechanism which powers much of the earth's geological activity.

Isostasy: the crust in balance

As page 13 explained, the upper parts of the continents are made of granitic crust (density 2.7 g/cm^3) and the lower parts are made of basaltic crust (density 3.0 g/cm^3). Beneath these layers is the upper mantle with a density of 3.4 g/cm^3. Note the difference in densities. As you might expect, the denser deeper layers provide support for the lighter materials above.

A similar arrangement is shown by the floating blocks in Fig. 2.22. Here pieces of wood (approx. density 0.65 g/cm^3) are supported by water (density 1.0 g/cm^3). Notice that the taller blocks are submerged deeper into the water. This is because they are heavier and therefore need more support from below. Now compare Fig. 2.22 with the section of the earth's crust shown in Fig. 2.16. There is certainly a similarity in overall shape. It seems that the highest continental areas (mountains) need to have a deeper zone of granitic crust beneath them in order to 'sink' further into the denser layers below and gain sufficient support. This explanation does not mean that the mantle is liquid or that continents actually float in the same way as wood blocks on water. It merely shows how the different heights of continental areas (and therefore different weights of rock) are supported and balanced from below.

The process of 'keeping a state of balance' in the crust is called **isostasy**. As yet, geologists are unsure exactly how it works, but it is vital in keeping the crust stable when events change the distribution of weight on the earth's surface. Consider these two examples.

1. It is known that as mountains are worn away by erosion the rocks beneath them rise slowly to 'balance the loss'. This is why the roots of some ancient mountain chains can now be seen at the earth's surface even though originally they may have been buried several kilometres deep. This process of uplift can be compared to cutting the top off one of the wooden blocks in the tank: it will rise and 'float' higher in the water because it needs less support from below.

2. A large thickness of sediment (sand, mud, etc.) can sometimes build up in a shallow sea without the sea becoming completely infilled. The reason for this is that the weight of the sediment causes the sea bed to sink slowly (like adding an extra layer to the top of one of the blocks). This subsidence means that the sea bed stays below water and the sediments can continue to build up.

Fig. 2.22 Wooden blocks of different heights but the same density are floated in a tank of water to demonstrate the effect of isostasy

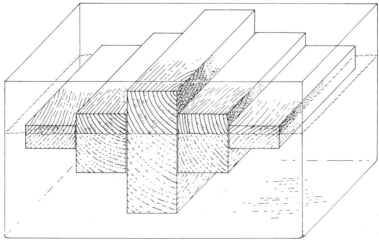

Questions

For questions 1–4, write the letters (A, B, C, ...) of the correct answers in your notebook.

1 The age of the earth is estimated to be
- A 500 000 yrs
- B 4 600 yrs
- C 300 000 000 yrs
- D 1000 million yrs
- E 4 600 000 000 yrs

2 The average density of the whole earth is
- A 5.5 g/cm^3
- B 2.7 g/cm^3
- C 5.5 tonnes/ mm^3
- D 3.0 g/mm^3
- E 1.0 g/cm^3

3 The Mohorovičić discontinuity is between
- A the inner and outer core
- B the crust and mantle
- C the outer core and mantle
- D the Nazca and Pacific plates
- E the oceanic and continental crust

4 Which **one** of the following plates is surrounded by subduction zones on all sides?
- A African Plate
- B Philippine Plate
- C Nazca Plate
- D Indo-Australian Plate
- E Pacific Plate

5 Use the following list to choose and write down the term which fits each description (a) – (g) below.
continental crust / mantle / inner core / asthenosphere / oceanic crust / atmosphere / outer core

(a) Zone made of nickel and iron in a solid state
(b) Zone with an average thickness of 6 km containing relatively young basaltic rocks
(c) Zone of gases surrounding the planet
(d) Zone of granitic rocks which geologists can most easily collect samples from
(e) Zone immediately below the lithosphere which is thought to allow plates to move over the deeper parts of the earth
(f) Zone believed to be in a liquid state and the cause of the earth's magnetic field
(g) The largest of the earth's internal zones

6 This diagram shows the pattern of shock waves recorded at a seismograph station after an earthquake.

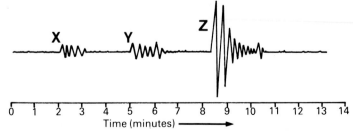

(a) Which point (X, Y or Z) marks the arrival of (i) the first S waves, (ii) the first L waves and (iii) the first P waves?

(b) How much time passed between the arrival of the first P wave and the first S wave?
(c) Assuming that P waves travel at 8km/s through the crust how far is this seismograph station from the epicentre. At what speed do S waves travel through the crust?
(d) Which type of earthquake waves
  (i) will not pass through the earth's outer core?
  (ii) cause the most damage to surface buildings?
  (iii) pass through all the earth's internal zones?
  (iv) have the lowest velocity?
(e) With the aid of a diagram describe the type of machine which records seismographs like that shown above.

7 The map shows some of the main plate boundaries on one part of the earth.

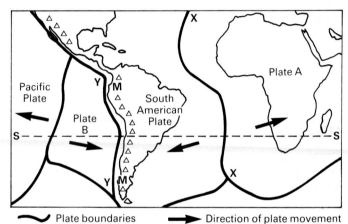

~ Plate boundaries   ➤ Direction of plate movement

(a) Name plate A, plate B and mountain range M.
(b) Line X–X is the Mid-Atlantic Ridge.
  (i) What type of lava is erupted along this ridge?
  (ii) Briefly explain how the eruption of this lava is related to the movement of plates on either side of the ridge.
(c) Line Y–Y is the site of a subduction zone where plate B descends beneath the South American continent.
  (i) What feature would you find on the ocean floor at the site of this subduction zone?
  (ii) Why is it impossible for the South American continent to descend into the subduction zone?
  (iii) Explain how subduction and plate movement have caused the formation of mountain chain M.
(d) How far (approximately) would you expect the plates shown on the map to be moving each year?
(e) Try to draw a cross-section along line S–S; this should be a 'side view' of the plates and continents which shows what is happening below the surface (if necessary you could use Fig. 2.21 as a guide but your diagram must relate to the area on the map above).

3 Minerals

It is a common mistake to confuse **minerals** and **rocks**. To avoid making this mistake, you need to learn what each of these terms means to a geologist.

What are minerals?

To understand about minerals we need a little knowledge of chemistry. Chemists often speak of **elements** (such as lead or sulphur) and **compounds** (such as lead sulphide). A compound is formed when two or more elements are combined by chemical bonding. A mineral is simply a solid chemical element or compound which is *formed naturally* in the earth. For example, the compound which chemists call lead sulphide is a common mineral which geologists know as **galena** (see Fig. 3.1).

There are several thousand different minerals known from the earth but only about thirty occur commonly. Every mineral can be given a chemical formula: for example, galena has the formula PbS (Pb lead, S sulphur).

What are rocks?

There is a great variety of rocks in the earth but the one thing they have in common is the fact they are made up of minerals. Some rocks, such as **meta-quartzite**, contain only one mineral (**quartz**), but most have a number of different minerals held together forming the solid rock.
*Study the photograph of the rock type **granite** (see Fig. 3.2 on page 33) and note the minerals it contains.*

When examining any rock the first questions a geologist asks are 'what minerals does it contain?' and 'how are these minerals held together?'. The answers will help in naming the rock type and understanding how it was formed.

Fig. 3.1 The mineral galena: note the cubic crystals

Remember

Elements	Example
	Silicon (Si) and **Oxygen** (O)
make up	make up
Minerals	**Quartz** (SiO_2)
make up	Quartz plus feldspar and mica make up
Rocks	**Granite**

Elements and minerals of the crust

Before studying rocks and minerals in detail it is worth learning something about the elements of which they are made. Over a hundred different elements are known to chemists but geologists normally consider these in two groups: the **major elements** which are present in large quantities in the earth's crust; and the **minor elements** which are much less common. These two groups form the basis for dividing minerals into the common **rock-forming minerals** (made of major elements) and the rarer **economic minerals** (made of minor elements).

Major elements and rock-forming minerals

There are eight elements in this group, and they are so common that they make up 99% of the weight of the earth's crust (see Fig. 3.3).

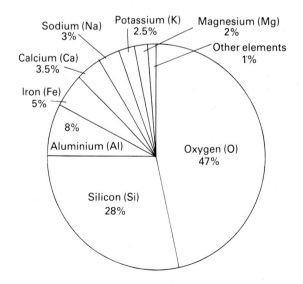

Fig. 3.3 The major elements occurring in the earth's crust (percentage by weight)

By far the greatest percentage of the crust is made up of silicon (Si) and oxygen (O). These two elements readily combine with each other, and with other elements, to form the commonest mineral group on earth: the **silicate minerals**. Silicates make up over 98% (by weight) of crustal rocks: for example, all the minerals in the granite shown on page 33 belong to this group. Although the silicates are the main rock-forming minerals, the **carbonates** (containing oxygen and carbon) are also important. The commonest carbonate is the mineral **calcite** of which the rock types limestone and marble are made.

Table 3.1 The commonest types of rock-forming minerals

Quartz*	Pure silicate (often called silica): formula SiO_2.
Feldspar family*	A range of silicates containing aluminium, sodium, potassium and calcium. Commonest types are **orthoclase feldspar** and **plagioclase feldspar**.
Mica family*	Complex silicates whose main types are **biotite mica** and **muscovite mica**. **Clay minerals** also belong to this group.
Amphibole family	Complex silicates, commonest type called **hornblende***.
Pyroxene family	Silicates rich in magnesium, iron and calcium. Commonest type called **augite**.
Olivine	Silicate of magnesium and iron: formula $(Mg,Fe)SiO_4$
Calcite	Not a silicate, but calcium carbonate: formula $CaCO_3$.

Remember: silicate means containing silicon and oxygen; carbonate means containing carbon and oxygen

* These minerals can be seen in Fig. 3.4 on page 33.

Table 3.2 Some important economic minerals

Mineral	Chemical formula	Elements extracted, and uses
Haematite	Fe_2O_3 ⎫	**Iron:** iron and steel goods (machines, vehicles, tools, girders, etc.)
Magnetite	Fe_3O_4 ⎭	
Pyrite	FeS_2	**Sulphur** (chemicals); iron cannot be economically processed
Chalcopyrite	$CuFeS_2$ ⎫	**Copper:** pipes, electrical cables, alloys, etc. Also copper chemicals used for dyes, etc.
Malachite	$CuCO_3Cu(OH)_2$ ⎭	
Galena	PbS	**Lead:** batteries, alloys and chemicals
Sphalerite	ZnS	**Zinc:** alloys, chemicals and galvanising
Cassiterite	SnO_2	**Tin:** alloys, chemicals and plating cans
Fluorite	CaF_2	Flux in steel industry; also **fluorine** chemicals
Barite	$BaSO_4$	**Barium:** chemicals
Halite (rock salt)	$NaCl$	Food industry, road salt; also **sodium** and **chlorine** chemicals
Gypsum	$CaSO_42H_2O$	Used to make plaster

The ore bodies of gypsum and rock salt actually take the form of layers of rock, but these minerals are not really common enough to be included in the rock-forming group.

Minor elements and economic minerals

The minor elements are rare and, in total, amount to only 1% of the weight of the crust. However, this group contains elements which are vital to industry. As the following figures show, some of the most sought after elements make up only a tiny proportion of the crust's weight.

Copper (Cu) 0.007%, Uranium (U) 0.0004%, Gold (Au) 0.000 000 5%

Minor elements are not present in large enough quantities to form gigantic masses of minerals and rocks like the silicates or carbonates. Instead these elements tend to be concentrated in certain places as fairly small 'pockets' of **economic minerals** (called **ore bodies**). To supply the world's needs, these 'pockets' have to be found and the minerals mined or quarried. If the present rate of mining and quarrying continues, however, the earth's supply of some of these elements may be in danger of running out.

Fig. 3.5a Derelict buildings at an ancient Derbyshire lead mine

Fig. 3.5b Large-scale modern opencast copper mining, Papua New Guinea

Crystals and mineral structure

Minerals form and grow by crystallisation. When this takes place the elements of which the mineral is made arrange themselves symmetrically and bond together in a definite pattern. Because of their regular internal structure, many minerals have a regular external crystal shape.

Here is a practical way of studying how a crystal may grow into a regular shape. For this you will need some alum powder. You can buy it in the chemist (alum is not actually a natural mineral but a chemical compound which suits this experiment). Gently warm about 50 ml of water then slowly add alum (stirring as you do so) until no more can be dissolved. You now have a *saturated solution* of alum and if you pour this onto a saucer small crystals will grow as the solution cools.

When crystallisation has completely finished, pour off any liquid and choose the crystal with the most regular shape. Allow this crystal to dry on blotting paper and then attach it to a piece of thread.

Make another saturated solution of alum and set up the equipment shown in Fig. 3.6. Do not place your crystal in the jar until the solution has cooled. Give the crystal several days to grow then remove and dry it.

What shape is the crystal? Would you have produced such a regular shape if the crystal had been placed against the inside edge of the jar?

What conditions are needed for symmetrical (equal in all directions) crystal growth?

To see the relationship between the internal arrangement of atoms and the external shape of a crystal you should look at some salt (NaCl) crystals under a magnifying glass. How is their shape related to the internal structure shown in Fig. 3.7?

Crystal systems

Geologists divide the wide range of crystal shapes into seven main types, or systems. For example, the mineral **galena** shown in Fig. 3.1 belongs to the **cubic crystal system**, while **quartz** (Fig. 3.4 on page 33) belongs to the **hexagonal crystal system**. Other minerals which show crystals of the cubic system include **halite, fluorite, pyrite** and **sphalerite**.

Within each of the seven systems a range of related shapes is possible. For example, Fig. 3.8 shows a few of the 'variations' which occur within the basic shapes of the cubic and hexagonal systems.

Fig. 3.6 Growing your own crystals

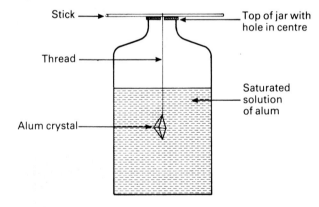

Fig. 3.7 The internal structure of salt

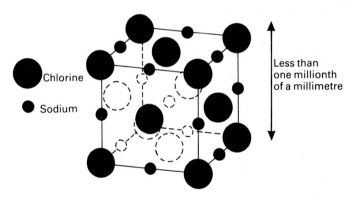

Fig. 3.8 Different crystal shapes within the (a) cubic and (b) hexagonal systems. Note that each flat regular surface is known as a crystal face

(a)

(b)

  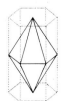

21

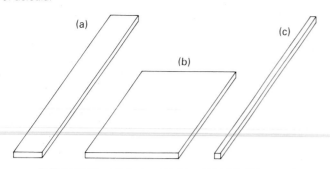

Fig. 3.9 Mineral habit: (a) bladed, (b) tabular, (c) 'needle-like' or acicular

Crystal habit

The term 'habit' is used to mean the overall shape of a crystal. Since there are so many variations within each crystal system it is often easier to describe a mineral's habit than to decide which system it belongs to. The following terms are used (see Fig. 3.9).

- **Tabular** habit: crystals with a flat plate-like or table-top shape, for example, **micas**.
- **Acicular** habit: long thin needle-like crystals.
- **Bladed** habit: long flat blade-shaped crystals.
- **Fibrous** habit: crystals that are thin, fine and hair like; **asbestos** (Fig. 3.10 on page 33) is so fibrous that it can be woven into cloth (for fireproof fabric).
- **Reniform** habit: the kidney shape of **haematite** (Fig. 3.11 on page 33).

How to identify minerals

Part of a geologist's basic training is learning how to identify minerals. Since each mineral type has its own chemical composition and internal structure, it will have a number of properties which are unique to it. The following pages explain how to discover these properties by observing and testing in a scientific way.

Observation 1: crystal shape

The shape of a mineral can be a useful aid to its identification. For example, you would normally expect galena to have cubic crystals. Use the previous sections of this chapter as a guide when you are observing and describing the shapes of mineral specimens.

Observation 2: mineral colour

Colour is an obvious feature but it will not always help you to identify a mineral. Many minerals show a range of colours. **Fluorite**, for example, can be colourless, green, yellow, purple or blue; and **quartz** shows a similar variation. Some colours result from tiny amounts of other elements 'contaminating' the mineral. For example, purple quartz (called amethyst) contains traces of titanium, while red quartz (jasper) contains iron.

Observation 3: mineral streak

'Streak' means the colour of the powdered mineral. It can be found by rubbing the specimen on a 'streak plate' (simply the back of a porcelain tile). This test is most useful for metallic minerals. For example, although the surface colour of **haematite** may vary, this mineral always gives a dark red streak. See Fig. 3.12 on page 33. However, this test is not very helpful for silicates and non-metallic minerals since they nearly all show a white streak.

Observation 4: mineral lustre

This describes the way light is reflected by a mineral. Several terms are used:

- **metallic** looks and shines like metal (example: **galena**);
- **vitreous** or **'glassy'** glistens like broken glass (example: **quartz**);
- **dull** no reflection at all (example: **malachite**).

While looking at lustre it is worth while also noting how light passes through the specimen. You can use one of these terms to describe what you see:

- **transparent** clearly see through it;
- **translucent** vaguely see light through it;
- **opaque** no light passes through.

Observation 5: mineral hardness

This property is tested by scratching the specimen (but do not ruin good crystals). **Mohs' scale of hardness** (1–10) arranges ten standard minerals in order from **talc** (the softest mineral known) to **diamond** (the hardest mineral known). Other minerals can be compared with minerals on Mohs' scale as follows. If a mineral can be scratched by fluorite (number 4), but will scratch calcite (number 3), then it must have a hardness of between 3 and 4.

There is actually no need to have a set of Mohs' minerals to test for hardness because common objects can be used as scratching tools (see Fig. 3.13). When testing for hardness always check your scratch through a hand lens. In some cases, what looks like a groove is only a coloured mark left by the instrument you tried to scratch with.

Fig. 3.13 Mohs' scale of hardness

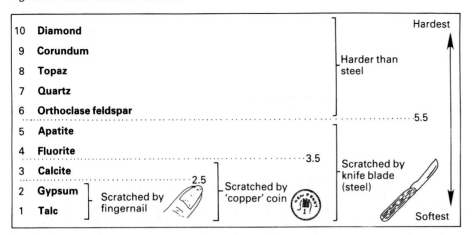

Fig. 3.14 Some examples of mineral cleavage

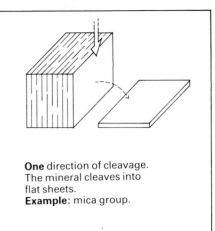

One direction of cleavage. The mineral cleaves into flat sheets. **Example:** mica group.

Observation 6: mineral cleavage and fracture

'Cleavage' means a plane of weakness along which a mineral or rock can be broken. In minerals these planes occur along lines of weaker bonds in their internal structure. As Fig. 3.14 illustrates, some minerals show only one set of cleavage planes while others have two, three or four sets.

Sometimes cleavage in a mineral can produce a new shape. For example, a cubic crystal of **fluorite** can be converted into a diamond shape when its corners are cleaved away (see Fig. 3.14). In other minerals, cleavage is parallel to the crystal faces. For example, a cubic **galena** crystal simply cleaves into smaller cubes.

Checking for cleavage needs a certain amount of care since cleaved surfaces can easily be confused with crystal faces. Begin your study by trying to cleave the minerals mentioned in Fig. 3.14 and comparing your results with the drawings. Try also to cleave **calcite** into the rhombic shapes shown in Fig. 3.15 on page 33.

With practice you will not need to break specimens when testing for cleavage; you will learn to recognise the faint parallel lines on the mineral surface which mark the position of cleavage planes within. A hand lens is useful for this.

Minerals with no cleavage will **fracture** unevenly. **Quartz**, for example, shows the **conchoidal** (shell-like) fracture illustrated by Fig. 3.16 on page 33.

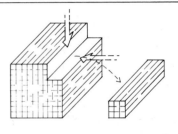

Two directions of cleavage. The mineral cleaves into long fragments. **Example:** feldspar group.

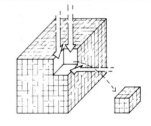

Three directions of cleavage. The mineral cleaves into pieces roughly cubic in shape. **Example:** halite and galena.

Observation 7: mineral reaction to acid

Testing a mineral specimen with a drop of dilute hydrochloric acid (HCl) may tell us something about its chemical composition. **Calcite** will fizz rapidly as bubbles of carbon dioxide (CO_2) are given off. Most sulphide minerals (particularly **galena**) will produce the unmistakable 'rotten eggs' smell of hydrogen sulphide (H_2S). Remember that even dilute acid can be dangerous so always use a dropper bottle and wipe up any acid left after you have finished testing.

Observation 8: mineral density or specific gravity

The normal way of calculating density is to use the formula:

$$\text{density} = \frac{\text{mass}}{\text{volume}}.$$

However, since it is difficult to find the volume of irregular-shaped

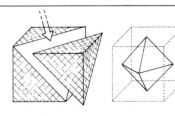

The cleavage of fluorite
Four directions of cleavage but none parallel to crystal faces, therefore all eight corners can be cleaved away to produce an octahedral (diamond) shape.

23

minerals, an alternative measurement called **specific gravity** (S.G.) is usually worked out. Fig. 3.17 explains how this is done.

Try this calculation for yourself.
Weight in air = 90 grams.
Weight in water = 78 grams.
What is the specific gravity of the mineral?
Which mineral has this specific gravity? (See the table opposite.)

When carrying out quick tests you do not need to calculate the S.G. accurately. With practice you will learn to judge which minerals feel heavy or light for their size.

Observation 9: mineral magnetism

This is a useful property for identifying **magnetite**, which is the only common mineral that deflects a compass needle.

Observation 10: mineral taste

This test identifies **halite** which tastes very salty. However, it is not a good idea to taste most other minerals, especially those which look metallic, or those that you have just tested with acid!

Quick revision check: identifying minerals	
Mineral shape	Very useful if mineral happens to be in the form of a good crystal
Mineral colour	Beware: results can be misleading
Mineral streak	Useful property for metallic minerals
Mineral lustre	Useful if you understand the terms properly
Mineral hardness	One of the best observations: always useful
Mineral cleavage and fracture	Very useful but only after practice
Mineral reaction to acid	Good for calcite and galena
Mineral specific gravity	Very useful if accurately calculated
Mineral magnetism	Good for magnetite
Mineral taste	Good for halite

With practice it becomes quite easy to make observations and tests on a mineral specimen. As you do this it is important to write down all your results. In this way you will build a list of all the properties typical of the particular mineral you are studying. You can then check this list with the identification table opposite and find the correct name for your mineral. Although this identification table uses mineral hardness as the first test, it is important to collect a full list of properties before trying to name a specimen.

In the field, you will probably rely on quick identification tests but if you are writing your results out fully

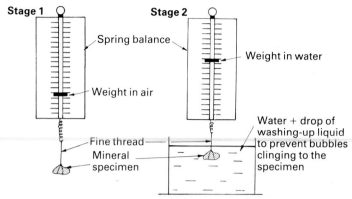

Fig. 3.17 How to calculate specific gravity

Stage 1 Attach specimen to balance and find weight in air.
Stage 2 Totally submerge specimen (still on balance) in water and find weight in water.
Stage 3
Work out specific gravity:

$$\text{S.G.} = \frac{\text{Weight in air}}{\text{Weight in air} - \text{Weight in water}}$$

(for example, in an examination) you should use the following style.

DESCRIPTION AND IDENTIFICATION OF MINERAL SPECIMEN X	
Crystal shape:	Some crystals show a definite cubic shape.
Colour:	Dull grey.
Streak:	Light grey.
Lustre:	Metallic.
Hardness:	Mineral cannot be scratched by a fingernail but can be scratched by a copper coin. This shows a hardness of between 2 and 3.5 on Mohs' scale.
Cleavage:	Mineral cleaves along planes parallel to the edges of its cubic crystals.
Acid:	Faint smell of hydrogen sulphide (H_2S) detected.
Density (S.G.):	No accurate measurement could be made but the specimen seemed heavy and dense.
CONCLUSION:	The information gathered above leads to the conclusion that Specimen X is a sample of GALENA (PbS).

Remember: Test first → then note results → conclude with a name.

More information about rock-forming minerals can be found in Chapters 5–8 which deal with rock types. See Chapter 15 for more information about economic minerals and ores.

Fig. 3.18 A complete kit for testing minerals

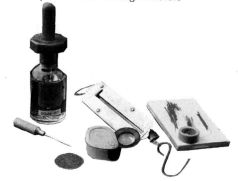

Mineral identification table

Scratch test	Hardness (Mohs)	Colour	Lustre	Streak	S.G. (Density)	Crystal system, habit, cleavage and fracture	Mineral	Other properties and comments
Scratched by fingernail	2	Colourless, white or tinted yellow	Very variable	White	2.3	May occur as platy crystals but often has no regular shape. Occasionally fibrous	Gypsum $CaSO_4 2H_2O$	So soft it is almost powdery*
Scratched by coin but not by fingernail	2½	Colourless to white; may be tinted pink/brown	Glassy	White	2.2	Belongs to cubic system. Cleaves into cubes	Halite NaCl	Has a salty taste*
	2½	Colourless but may look silvery	Glassy	White	3	Tabular crystals which easily flake due to excellent cleavage*	Muscovite mica (Silicate) R.F.	Crystals of these minerals are slightly flexible
	2½	Black to dark brown	Glassy	White	3	Tabular crystals which easily flake due to excellent cleavage*	Biotite mica (Silicate) R.F.	
	2½	Lead-grey	Metallic	Lead-grey	7.5*	Belongs to cubic system. Cleaves into cubes	Galena PbS	Gives off foul smell with HCL
	3	Colourless or white	Normally glassy	White	2.7	Shows a wide range of crystal forms and habits. Cleaves into 'rhombs'	Calcite $CaCO_3$ R.F.	Fizzes with * dilute HCl
	3 to 3½	Colourless, white or tinged pink	Variable	White	4.5*	Shows a range of forms and habits. Two cleavage planes	Barite $BaSO_4$	Very dense yet looks non-metallic. Also called barytes
	3½	Bright green*	Dull	Very pale green	4	Rarely shows good crystal form. May appear banded	Malachite $CuCO_3Cu(OH)_2$	Very distinctive colour
Scratched by steel blade but not by coin	4	Very variable: commonly white purple or yellow	Glassy	White	3.1	Belongs to cubic system. Cleaves into diamond shapes*	Fluorite CaF_2	Also called fluorspar
	4	Black to dark yellowish brown	Shines like resin*	White to faint brown	4	Often shows tetrahedral shape. Cleaves quite well.	Sphalerite ZnS	Also called zinc blende
	4	Yellow, like brass*	Metallic but tarnishes	Faint green/black	4.3	True crystal shape is square prism, but rarely well formed	Chalcopyrite $CuFeS_2$	Looks like pyrite but is softer*
Not scratched by steel blade	5½	Dark green to black	Glassy	Very light grey	3.2	Good crystals are rare. Two cleavages at 120°	Hornblende (Silicate) R.F.	Very difficult to distinguish from each other and olivine (mafic minerals)
	5½	Dark green to black	Glassy	Very light grey	3.4	Crystals often show 8-sided form. Two cleavages at 90°	Augite (Silicate) R.F.	
	6	White to pink	Glassy	White	2.6	Usually has short flattened crystals which cleave in two directions	Orthoclase feldspar (Silicate) R.F.	Very common minerals in igneous rocks
	6	White to grey	Glassy	White	2.7	Similar shape and cleavage to orthoclase but often twinned	Plagioclase feldspar (Silicate) R.F.	
	6 (varies)	Greyish red to black	Dull to metallic	Dark red*	5.2	Sometimes occurs as 'kidney ore' (see Figure 3.11, page 33)	Haematite Fe_2O_3	Some earthy varieties softer than hardness 6
	6	Iron-black	Metallic	Black	5.2	Often found as shapeless masses. Very poor cleavage	Magnetite Fe_3O_4	Magnetic properties*
	6½	Looks like gold*	Metallic	Faintly greenish black	5	Belongs to cubic system. Crystal faces often show faint parallel lines	Pyrite FeS_2	Also called 'fools gold'*
	6½	Deep brown or black	Brilliant	White or pale yellow	7*	Crystals are usually square prisms. Often twinned.	Cassiterite SnO_2	Extremely dense*
	6½	Dark olive green or brown	Glassy	white to very pale grey	3.8	May show rectangular prisms but good crystals are rare. No cleavage.	Olivine (Silicate) R.F.	Looks similar to augite and hornblende (mafic mineral)
	7*'	Variable but colourless to milky white varieties are most common	Glassy	White	2.7	Shows many forms but commonly in hexagonal prisms. Conchoidal fracture	Quartz SiO_2 R.F.	Looks like grains of glass in rocks

* Very useful property
R.F. Rock forming (be prepared to identify small specimens within rocks)

Questions

For questions 1–10, write the letters (A, B, C, ...) of the correct answers in your notebook.

1 The two most abundant elements in the earth's crust are
A Ni and Fe   B Si and Mg   C Si and O
D Al and O   E Si and Al

2 Which **one** of the following is **not** a silicate mineral?
A quartz   B augite   C calcite
D olivine   E biotite mica

3 Which **one** of the following minerals could be scratched by a steel blade but not by a copper coin?
A galena   B quartz   C halite
D fluorite   E corundum

4 Which **one** of the following minerals contains **both** iron and copper in its atomic structure?
A chalcopyrite   B magnetite   C pyrite
D cassiterite   E haematite

5 Which **two** of the following minerals commonly form cubic crystals?
A hornblende   B halite   C galena
D gypsum   E plagioclase feldspar

6 Which **two** of these terms could be used to describe the shape (habit) of this crystal?

A tabular   B hexagonal   C prismatic
D reniform   E acicular

7 If a mineral specimen has a weight (in air) of 240 g and a weight (in water) of 160 g, then its specific gravity must be?
A 6   B 3   C 1.5
D 0.66   E 4.2

8 Which **one** of the following minerals could be foolishly mistaken for gold?
A cassiterite   B calcite   C orthoclase
D pyrite   E magnetite   feldspar

9 Which **three** of the following minerals are sulphides?
A pyrite   B gypsum   C sphalerite
D galena   E cassiterite

10 Which **two** of the following minerals are silicates with hardness of above 6 on Mohs' scale?
A olivine   B barite   C fluorite
D quartz   E cassiterite

11 The table below lists properties of four minerals collected from the waste heaps of an old mine.

Mineral	Colour/lustre	Hardness	Other properties
A	White	Scratched by copper coin	Cleaves into rhombs, reacts with dil. HCl
B	Silver-grey, metallic	Scratched by mineral A	Cleaves easily into cubes
C	Dark brown, resinous	Harder than copper coin	Cubic crystals
D	Pinkish-white	Same hardness as mineral A	Denser than minerals A and C but less dense than B

(a) A specimen of mineral B had a mass of 375 g and a volume of 50 cm³; what is the density of this mineral?
(b) Name minerals A, B, C and D.
(c) Give the chemical composition of minerals A and D.
(d) State one economic use for each of minerals B, C and D.
(e) Which one of the minerals could be classified as a rock-forming mineral? Name a rock made up of this mineral.
(f) How many sets of cleavage planes does mineral A have?

12 Imagine that you have been given specimens of quartz, gypsum and fluorite which are all white in colour. Explain the tests and observations that you would make (stating the results you would expect) if asked to find out which specimen was which.

13 Using only a streak plate and a copper coin explain how you would distinguish these five minerals from each other: haematite, halite, malachite, magnetite, fluorite.

14 Study the diagrams of mineral crystals. Note that the faint lines (where shown) represent traces of cleavage planes.

A    B

C 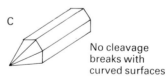   D

No cleavage breaks with curved surfaces

(a) For each of the minerals A to D:
(i) name the crystal shape (or habit) shown
(ii) describe the type of cleavage (or fracture) shown
(iii) name the mineral illustrated.
(b) Which two of these minerals would you expect to be present in granite?
(c) Explain why minerals develop a particular crystal shape and cleavage pattern.

4 An introduction to rocks

Now that you know something about minerals, we can go on to study the rocks which are formed from them. Rocks can be divided into three main groups:
1. **igneous rocks,**
2. **sedimentary rocks,**
3. **metamorphic rocks.**

This division is based on the way in which the rocks have been formed.

Igneous rocks

Definition

Igneous rocks are formed when **magma** cools and solidifies.
The name comes from the Latin word 'ignis', meaning fire.

Explanation

- Magma is molten rock.
- Magma is formed in the crust and upper mantle where there is sufficient heat to partially or completely melt rocks.
- Once formed, magma tends to rise upwards. If it reaches the earth's surface it is then called **lava**.
- As magma cools, mineral crystals begin to grow (especially crystals of silicate minerals).
- In most cases the crystals grow and interlock, forming a hard **crystalline rock**. But if the magma was erupted into the air as a spray of droplets it cools rapidly and falls to earth as particles of **volcanic ash**.

Examples

Granite, gabbro, rhyolite, dolerite, basalt, pumice, obsidian, agglomerate and volcanic ash (tuff).

Main text reference

Igneous rocks are dealt with on pages 31 to 41.

Sedimentary rocks

Definition

Sedimentary rocks are formed from **deposits** of material which has originally come from older rocks or living organisms.
The name comes from the Latin word 'sedo', meaning to settle or sit down.

Explanation

- All rocks exposed at the earth's surface are **denuded** (worn away) by **weathering** and **erosion**.
- The material worn away from these rocks is **transported** (carried away) by gravity, wind, ice, sea and rivers.
- The transported material may be **fragments** of rocks and minerals (for example, pebbles and sand grains), or it may be **dissolved** in water.
- Eventually the transported material is **deposited** (laid down) as a layer or **bed** of **sediment** (for example, in a desert or on a sea shore).
- The deposition may involve the **precipitation** of dissolved material (for example, a bed of salt left by the evaporation of sea water).
- Some beds of sediment are made from the remains of living organisms (for example, coal formed from trees).
- The beds of sediment, however formed, are slowly **compacted** (pressed together) as other material continues to be deposited above. Eventually the separate pieces of loose sediment become **lithified** (joined together) into a solid sedimentary rock (lithified literally means 'turned to stone').

Examples

Conglomerate, breccia, sandstone, mudstone, clay, shale, coal, limestone, chalk, gypsum and rock salt.

Main text reference

Sedimentary rocks are dealt with on pages 64 to 79.

Metamorphic rocks

Definition

Metamorphic rocks are formed when **heat** and/or **pressure** cause rocks to **change**.
The name comes from the Greek words 'meta', meaning change, and 'morphe', meaning form.

Explanation

- As conditions within the earth change, rocks are also forced to change.
- Heat and pressure may cause the chemical elements in the original rock to **react** and **re-form** into new minerals.
- The new metamorphic minerals are often **aligned** (arranged) in a parallel pattern because they have formed under pressure.

- During metamorphism no elements are taken away from the original rock and no elements are added; the original elements are simply rearranged to suit new conditions.
- Although metamorphism may involve great heat, the rock **does not melt** as it changes; if it did, a magma would be formed and the new rock would be classified as igneous.
- Sedimentary rocks are most likely to be affected by metamorphism, because they were originally formed at fairly low temperatures and pressures. Since igneous rocks were originally formed by heat, they do not change greatly if re-heated.

Examples

(The name in brackets is the original rock which has been changed.)

Marble (limestone); metaquartzite (sandstone); slate, schist, gneiss and hornfels (all from fine sediments such as mudstone and shale).

Main text reference

Metamorphic rocks are dealt with on pages 80 to 87.

The rock cycle

As Fig. 4.1 shows, the earth's crust is made up of all three major rock types. It follows that the whole crust has actually been formed by the processes which produce igneous, sedimentary and metamorphic rocks. If you look again at the definitions of these rocks types you will see that only igneous rocks are formed from new material brought into the crust. This 'new' material is magma which rises from the mantle below.

Both sedimentary and metamorphic rocks could be described as being made of 'second-hand' materials. The materials in a sedimentary rock originally belonged to other rocks which were broken down by weathering and erosion. The materials in a metamorphic rock originally belonged to other rocks which were forced to change under pressure and/or heat. The original crust of the earth must have been made entirely of igneous rocks. The other rock types have developed as the crust has changed through time.

This changing and re-forming of rock materials continues all the time. It is called the **rock cycle** (see Fig. 4.2). Study it for yourself and work out how materials from any one of the rock types can be re-formed into either of the other two types.

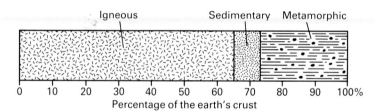

Percentage of the earth's crust

Fig. 4.1 The main rock types in the earth's crust. These figures are for the *whole* crust. The most common type at the surface is sedimentary which forms a relatively thin 'covering' over the deeper igneous and metamorphic types

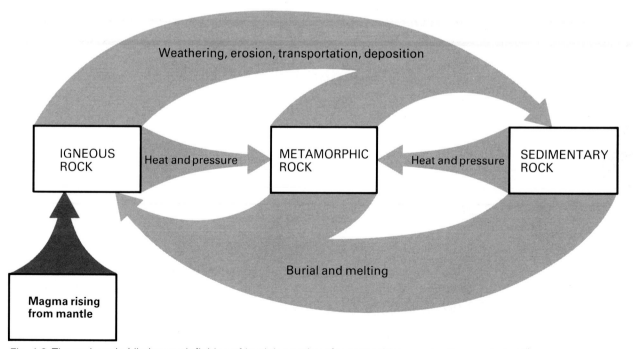

Fig. 4.2 The rock cycle (dictionary definition of 'cycle': a series of events which constantly repeats itself)

Studying rocks

When you first begin to study rocks there seems to be a bewildering variety of types, all with different names to learn. However, things become easier when you realise that only one general method is used to describe and identify all rocks types, whether **igneous, sedimentary** or **metamorphic**. This method is based on the fact that rocks are made of minerals, and considers these minerals in two ways.

1. What minerals are present in the rock and in what proportions do they occur? This is called the **mineralogy of the rock**.

2. What size and shape are the minerals and how are they arranged and held together in the rock? This is called the **texture of the rock**.

Studying the **mineralogy of the rock** mainly involves identifying the rock-forming minerals of which it is made. The rock's colour is related to its mineralogy since it depends on the colours of the minerals within. Rock mineralogy will be dealt with further in each of the separate chapters on igneous, sedimentary and metamorphic rocks.

Studying the **texture of the rock** will tell you how the minerals came together to form that rock. The three main types of rock have all formed in different ways so they all show different textures. Table 4.1 illustrates and explains the more common textures, but you may need a hand lens to see some of these features in real specimens. Rock texture also governs the feel (for example, rough, smooth, gritty, etc.) and strength (for example, breaks easily, breaks into layers, breaks with jagged surfaces, etc.) of a specimen.

Questions

1 What is magma? In what situations can magma be called lava?

2 Rearrange these terms into the correct order so that they list the processes which form sedimentary rocks.
lithifaction / denudation / transportation / deposition

3 Briefly explain what causes metamorphism. Why are sedimentary rocks more likely to be metamorphosed than igneous rocks?

4 Sort these rock types into three lists: one for igneous rocks, one for sedimentary rocks, and one for metamorphic rocks.
marble / conglomerate / shale / tuff / pumice basalt / hornfels / chalk / rhyolite / schist granite / coal / metaquartzite / sandstone / slate / limestone / gneiss / dolerite

5 With the aid of diagrams and examples, explain the difference between a crystalline texture and a fragmental texture. What can be learnt by the study of rock textures?

	Igneous	Sedimentary	Metamorphic
Mineral size and Mineral shape	Depends mainly on the rate at which the magma cooled and crystallised. Rapid cooling results in the growth of small irregular-shaped crystals. Slow cooling allows time for larger well-formed crystals to grow.	Depends mainly on the conditions of deposition and the style of transportation that has delivered the sediment. Long distances of transportation and/or deposition in calm conditions tend to produce fine-grained sediments (e.g. muds laid down in offshore waters). Shorter distances of transportation and/or deposition in turbulent conditions tend to produce coarse-grained sediments (e.g. pebbles laid down on beaches).	Depends mainly on the original rock type and the amount of heat and/or pressure that has caused the metamorphic changes. Relatively low pressures and/or temperatures cause little change to the texture of the original rock; only small poorly developed new metamorphic minerals are formed. High pressures and/or temperatures produce widespread changes to the original rock texture; allowing larger well-formed new metamorphic minerals to crystallise.
Arrangement of minerals (This is illustrated by diagrams showing typical rock types under a microscope. Each rectangle shows an area of 2 cm by 1 cm of the actual rock)	In general, igneous rocks show a **crystalline texture** with mineral crystals interlocked like a mosaic. Because crystals were free to grow in any direction they have a **random orientation.** Quartz Feldspar Biotite mica Example shows a **granite** (page 40) with interlocking crystals of quartz, feldspar and biotite mica.	In general, sedimentary rocks show a **fragmental texture** with individual mineral particles packed together and held by a **matrix** or **cement**. The particles may show some **alignment** because they were deposited in a layer or bed. Quartz Example shows a **sandstone** (page 68) with fragments of quartz (sand grains) held together by a cement of calcite. This cement has precipitated to fill the spaces between the sand grains and holds the sediment together as a solid rock.	In general, metamorphic rocks show a **crystalline texture** although the amount of crystallisation depends on how intense the metamorphism was. Because they developed under pressure many minerals are **aligned** in a particular direction. Garnet Mica Example shows a **schist** (page 84) with crystals of mica aligned and arranged around a crystal of garnet. Both mineral types have grown as a result of metamorphism and the small dots represent all that is left of the fine particles of the original shale (shale is a sedimentary rock).

Table 4.1 Rock types: textures and origins

5 Igneous rocks and volcanoes

Before beginning this chapter re-read page 27 to remind yourself of the main facts about igneous rocks. Write short definitions of the terms igneous, magma, lava, crystalline rock, and eruption.

Igneous rocks can be divided into two main groups.
1. **Extrusive igneous rocks** are formed from magma (that is, lava) which has solidified on the earth's surface. Another name for this group is **volcanic rocks**.
2. **Intrusive igneous rocks** are formed from magma which has solidified within the earth's crust.

Extrusive igneous rocks: an introduction to volcanoes

There are about 540 **active** volcanoes on earth, including about 80 known examples on the sea bed. The term 'active' is normally used for volcanoes which have erupted within living memory (in the last 80 years or so). Other volcanoes are described as **dormant** (resting) or **extinct** (dead), but these names should be used with care. Volcanoes which were assumed to be extinct have sometimes returned to life with disastrous results.

The distribution of active volcanoes was shown in Fig. 2.17 (page 14). As explained in Chapter 2, active volcanoes mainly occur at **plate margins**. This is because the energy associated with plate movement generates enough heat to melt rocks. At present, places such as the British Isles are well away from these active plate margins, so they do not suffer volcanic activity. However, if we study rocks formed in the past we realise that this has not always been the case. For example, rock found in the Lake District and North Wales tells us that huge volcanoes were erupting there about 450 to 500 million years ago. More recently (about 50 to 65 million years ago) lavas flowed over wide areas of what are now the islands of north west Scotland. Evidence such as this suggests that Britain (along with the rest of Western Europe) must have been in contact with plate margins of the past.

Fig. 5.1 A volcanic eruption on White Island, New Zealand. *Look at Fig. 2.18; which plate margin is this volcano near?*

As Fig. 5.2 shows, volcanoes erupt igneous material in three forms: solid, liquid and gas. We shall begin by studying liquid lava. The type of lava produced and the way it flows determine the shape of the volcano and the way it erupts.

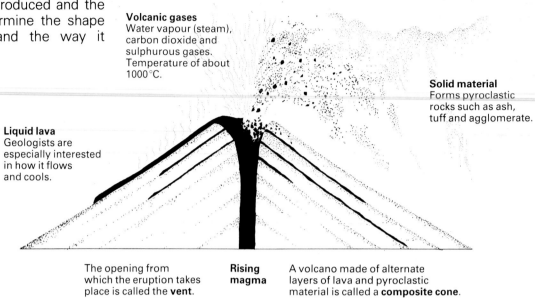

Fig. 5.2 Volcanic eruption

Volcanic gases
Water vapour (steam), carbon dioxide and sulphurous gases. Temperature of about 1000°C.

Solid material
Forms pyroclastic rocks such as ash, tuff and agglomerate.

Liquid lava
Geologists are especially interested in how it flows and cools.

The opening from which the eruption takes place is called the **vent**.

Rising magma

A volcano made of alternate layers of lava and pyroclastic material is called a **composite cone**.

Free-flowing mobile lava

This type of lava cools to form the rock known as **basalt**. Because it is free-flowing it moves easily from the vent and spreads to produce large broad cones called **shield volcanoes** (see Fig. 5.3). Mauna Loa in the Hawaiian Islands is an example of such a cone. It rises almost 10 km from the Pacific floor and, at its base, is big enough to cover an area the size of Wales.

Volcanic gases can escape easily through mobile lava, so intense pressure rarely builds up. Also, because it flows so freely, basalt lava moves away from the vent rather than solidifying there and forming a 'plug'. As a result of these features, shield volcanoes tend to erupt more often but with less violence than other volcanic types.

Slow-moving viscous lava

This lava produces the rock type known as **rhyolite**. Because it has a different chemical composition to basalt it flows much more slowly and tends to build narrow steep-sided cones (see Fig. 5.4).

Rhyolite lavas often solidify in the vent forming a 'plug' which causes pressure to build up beneath. This means eruptions can be extremely violent, with a mass of lava droplets and shattered vent rock being blasted out in a jet of volcanic gas. A devastating mixture such as this may travel rapidly down the side of the volcano as a **nuée ardente** (the name means 'glowing cloud'). When Mont Pelée on the island of Martinique (West Indies) erupted in 1902, it produced a **nuée ardente** which swept into the nearby town of Saint Pierre and wiped out all 30 000 inhabitants within seconds.

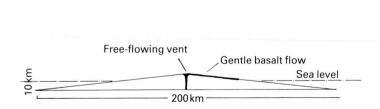

Fig. 5.3 Mauna Loa (shield volcano)

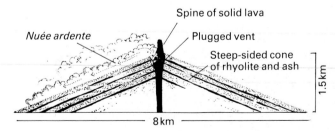

Fig. 5.4 Mont Pelée (steep-sided volcano)

It is impossible to classify all volcanoes as being like either Mauna Loa or Mont Pelée. There is a complete range of types with shapes and eruption styles between these two extremes

32

Fig. 2.19 The Atlantic ocean floor

Fig. 2.20 The Pacific ocean floor

Fig. 3.10 The fibrous habit of asbestos

Refer to these pictures when reading about plate tectonics in Chapters 2 and 13. *Try to recognise oceanic ridges, island arcs, trenches, and transform faults.*

Fig. 3.4 Specimens of silicate minerals
a muscovite mica, b biotite mica,
c quartz, d hornblende, e plagioclase
feldspar, f orthoclase feldspar

Fig. 3.2 The rock type granite is made up of quartz, feldspar and mica. *Can you recognise them?*

5 cm

Fig. 3.11 The reniform habit of haematite: often called 'kidney ore'

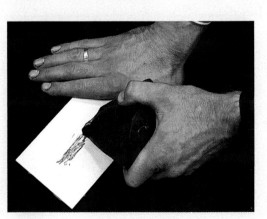

Fig. 3.12 Testing a mineral for streak

5 cm

Fig. 3.15 Calcite showing the three sets of cleavage planes which produce the 'rhombic'-shape fragments

5 cm

Fig. 3.16 Quartz showing conchoidal fracture

33

Fig. 5.6 Pillow lava, formed over 450 million years ago, now exposed at Strumble Head, Dyfed

Fig. 5.7 Specimen of vesicular lava

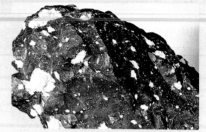

Fig. 5.8 Specimen of amygdaloidal lava

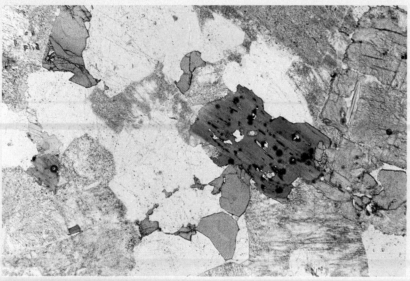

Fig. 5.17 Photomicrograph of granite. Note the crystalline texture of quartz (clear), feldspar (buff) and biotite mica (brown). The area in view is about 5 mm wide

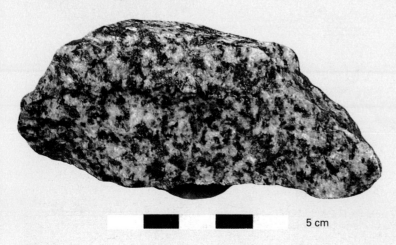

5 cm

Fig. 5.18 Specimen of gabbro: see page 41 for a full description

Fig. 6.4 Exfoliation

Features of lava flows

When lava has solidified into rock it usually shows features which are a result of how it flowed and cooled.

'Ropey' lava. This is a surface feature shown by free-flowing basalt lava. When the surface cools it forms a 'skin' which then becomes wrinkled by lava continuing to move beneath (see Fig. 5.5).

'Blocky' lava. This is another surface feature of basalt flows. It is produced by volcanic gas breaking through the skin and separating it into jagged blocks.

Flow-banded lava. This is found in viscous rhyolite flows. As the partly solidified lava moves, small crystals of the minerals become lined up in the direction of flow.

Pillow lava. When lava flows into water a thick skin is quickly formed. This skin then fills with lava (like a kind of blister) to produce a pillow-shaped mass. Pillows can separate from the flow and roll together forming a structure like that shown in Fig. 5.6.

Vesicular lava When volcanic gas is trapped during cooling, small round cavities (bubbles) are left in the rock (see Fig. 5.7). These are called **vesicles**.

Amygdaloidal lava (see Fig. 5.8) After vesicular lava has cooled, water may seep through the vesicles. If this water is carrying minerals in solution, **precipitation** (settling out from the solution) of minerals may take place and infill the vesicles. Vesicles infilled with minerals are called **amygdales**.

Flood basalts. These are produced by free-flowing lava erupting from long cracks or **fissures**. This type of flow mainly happens on the sea bed. It forms new basaltic crust along mid-oceanic ridges where two plates are moving apart (see page 15). On land, repeated fissure eruptions can 'flood' a wide area with

Fig. 5.5 'Ropey' lava, Canary Islands

layer upon layer of basalt. For example, the Deccan Plateau of India contains over 500 000 km^2 of basalt and, on a smaller scale, the Antrim basalt flows of Northern Ireland contain almost 4000 km^2.

Columnar jointing. This is formed during the cooling and contraction of thick flows (usually flood basalts which have been erupted on land). The contraction causes cracks to develop and the rock separates into a large number of vertical (usually six-sided) columns (see Fig. 5.9).

Fig. 5.9 Columnar jointing in basalt on the Isle of Staffa (near Mull), Scotland

Volcanic gases

These gases are difficult to study because, once erupted, they mix with each other and the air. Also it is not easy to find a geologist willing to collect samples from an active vent! The main gases are water vapour, carbon dioxide and sulphurous gases (Fig. 5.2). The presence of gas has important effects on other aspects of volcanic action. For example, gas pressure causes the eruption of a *nuée ardente* (page 32), and gas propels particles of volcanic ash into the air. Gas can also affect the lava as it solidifies. **Pumice** is a volcanic rock formed from lava which was turned to a froth by escaping gas. It is so full of vesicles (bubbles) that it will float (Fig. 5.11).

Fig. 5.11 A piece of (a) pumice shown with a specimen of (b) basalt of equal weight

(a)　　5 cm　　(b)

Fig. 5.10 'It is not easy to find a geologist willing to collect gas samples from a volcanic vent'

Fig. 5.12 Pyroclastic 'fall out'

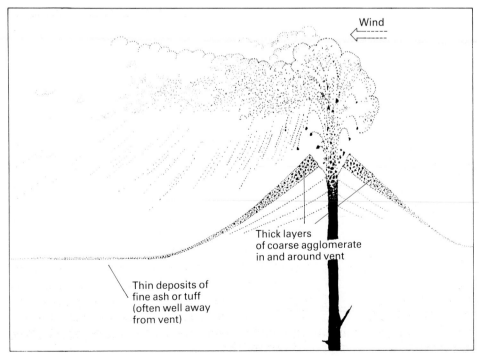

Wind

Thick layers of coarse agglomerate in and around vent

Thin deposits of fine ash or tuff (often well away from vent)

Pyroclastic rocks

The word 'pyroclastic' comes from Greek words meaning 'fire' and 'broken'. Pyroclastic rocks are formed from two kinds of 'fragments' blown out by volcanic eruptions:

- droplets of lava which solidify as they travel through the air,
- solid pieces of rock (from in and around the vent) which are shattered and blasted out.

The distance these fragments travel depends on their size and the force of the eruption which produced them. Fig. 5.12 illustrates the general rule that 'the finer the fragments, the further they travel'. Different types of pyroclastic rock are formed depending on the size of the fragments.

Fig. 5.13 Volcanic bombs. These examples are 8–12 cm in length but size can vary greatly

Examples of volcanic eruptions

The 1980 eruption of Mount St Helens in the Cascade Mountains of the U.S.A. was studied in detail. It produced a great ash cloud which covered a wide area. Close to the vent this cloud is known to have travelled at over 200 km/hour destroying everything in its path. Melted snow and mudflows also poured down the mountain sides adding to the chaos which killed at least 61 people.

The total amount of ash from Mount St Helens was estimated at 1 km^3. But this is small compared with the 18 km^3 of ash produced from the eruption of Krakatau (near the island of Java) in 1883. The blast from Krakatau was heard over 4000 km away in Australia. Over 35 000 people were killed by the tidal wave (*tsunami*) caused by this explosion, and floating pumice was a danger to ships for months.

Hot springs and geysers

These are found where volcanic activity is decreasing but the rocks below ground are still hot enough to heat any water that seeps through them. A **hot spring** is a place where this hot water flows back to the surface. It often contains dissolved minerals which are said to be good for health. A **geyser** is a more violent type of hot spring. What happens is that the water becomes heated to boiling point, and steam is formed. Steam pressure builds up under the ground until it shoots a jet of nearly boiling water up to the surface and into the air. An example is the 'Old Faithful Geyser' in the U.S.A. which 'performs' on average once every hour.

Hot spring water is used to provide heating or power in countries such as Iceland and New Zealand.

Agglomerate. This is made of the largest fragments and is found closest to the vent. It often contains fragments called **volcanic bombs**. They were originally large 'blobs' of lava which cooled into a rounded shape as they flew through the air (see Fig. 5.13)

Ash and tuff. The finer fragments fall to earth as volcanic ash. Eventually they may become compacted (pressed) together to form the solid rock known as **tuff**. Ashes and tuffs often show features such as **bedding** (different layers) which make them look like sedimentary rocks. This happens especially when ash falls onto the sea and then sinks to be deposited just like silt or any other normal sediment. A different sort of tuff occurs when the fragments are so hot that they fuse together after landing on the ground. This is called **welded tuff** and tends to be produced by *nuée ardente* eruptions.

Some volcanoes have cones made almost completely of ash and agglomerate.

Dust. This is the finest volcanic material. It can be carried and spread over great distances by the wind. Sometimes volcanic dust rises so high into the atmosphere that the weather is affected for weeks.

Intrusive igneous rocks

These rocks are produced when magma cools and crystallises within the earth. They are called **intrusive** because they push their way into the existing crustal rocks. Since this occurs well below the surface, geologists can never actually see intrusive rocks forming. Instead they have to study them many years later when they appear at the surface after the rocks above them have been worn away.

A number of different names are given to intrusions of igneous rock depending on their shape and their relationship with the surrounding **country rock** (the original rocks of an area that magma has intruded into). The three main types of intrusion are **dykes, sills** and **batholiths**.

Dykes

A dyke is formed when magma has intruded into a fracture or fissure through the original country rock, and solidified. As a result, dykes tend to be vertical (or nearly vertical) walls of igneous rock which cut right across the original pattern (for example, the **bedding**) of country rock (see Fig. 5.14). The width of dykes varies from a few centimetres to several kilometres, but most are a few metres across. The best examples in Britain are in the islands of north west Scotland where many dykes of a rock type known as **dolerite** stand out as small ridges across the land surface.

Fig. 5.14 Sill and dyke. Note: lava flows have a baked margin below their base but not above their top surface. Why is this?

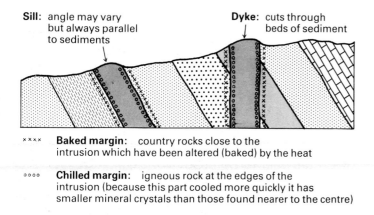

Sill: angle may vary but always parallel to sediments

Dyke: cuts through beds of sediment

× × × × **Baked margin:** country rocks close to the intrusion which have been altered (baked) by the heat

∘∘∘∘ **Chilled margin:** igneous rock at the edges of the intrusion (because this part cooled more quickly it has smaller mineral crystals than those found nearer to the centre)

Fig. 5.15 The batholith below south west England

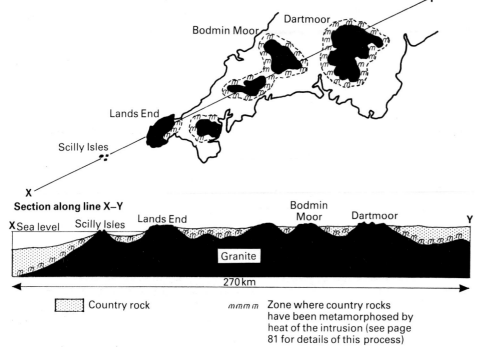

Section along line X–Y

Country rock

mmmm Zone where country rocks have been metamorphosed by heat of the intrusion (see page 81 for details of this process)

Sills

A sill occurs where magma intrudes between layers (beds) of sedimentary rock. Often sills are horizontal like the original layers of rock. But, if the layers of rock become **folded** (see page 93), the angle of the layer boundaries (and the sill) changes. As Fig. 5.14 shows, a sill may eventually be exposed at a folded (non-horizontal) angle. The fact that the angle of a sill varies should not confuse you if you remember that a sill is always parallel to the sedimentary beds. A large sill lies within the sedimentary rocks of the Northern Pennines and Northumberland. It is known as the Whin Sill and, because it is made of hard dolerite, it stands out as steep crags wherever it is exposed.

Batholiths

A batholith is a major intrusion where a huge mass of magma has invaded and/or replaced a vast amount of other rock. Batholiths have a shape best described as an 'upturned boat' and can be hundreds of kilometres long by tens of kilometres wide. Because enormous amounts of heat energy are needed to make intrusions of this size, most batholiths form in regions where the forces of mountain building are acting. As Fig 5.15 shows, a batholith of granite lies beneath south west England. This formed about 275 million years ago when plate movement caused mountains to be produced across much of Europe. Despite its age, only the upper parts (called **stocks**) of the south west batholith have so far been exposed at the surface. They stand out as areas of high moorland. Similar granite batholiths lie beneath other parts of Britain, for example, the Lake District and Northern Pennines.

Effects of igneous intrusions

The heat from an intrusion can cause **metamorphism** of the surrounding country rocks, particularly around large intrusions which have a lot of heat to lose.

Watery fluids which are rich in minerals may escape from large bodies of magma and seep through the surrounding rocks to form **mineral veins**.

Identifying igneous rocks

As explained on page 29, the identification of all types of rock depends on studying two things.
1. **Texture**: what sizes and shapes are the mineral grains and how are they arranged in the rock?
2. **Mineralogy**: what minerals are present and in what proportions are they found?

Texture of igneous rocks

The texture of an igneous rock depends mainly on the way it cooled and solidified. Most igneous rocks have a **crystalline texture** produced by minerals growing in a cooling magma until they eventually form a mosaic of interlocking crystals (see Table 4.1, page 30 and Fig. 5.17 on page 34). The size of these crystals will depend on the rate of cooling and, as a general rule, you should remember that 'the slower the cooling the larger the crystals become'. Three different textures are recognised in crystalline igneous rocks, depending on crystal size:

- **coarse grained**, where the individual mineral crystals can be clearly seen by eye (examples: **gabbro** and **granite**);
- **medium grained**, where the rock has a speckled appearance but the separate crystals can only be seen properly with the aid of a hand lens (example: **dolerite**);
- **fine grained**, where the specimen has a practically uniform colour and the separate crystals can only be distinguished under a microscope (examples: **basalt** and **rhyolite**).

Because crystal size depends on rate of cooling it also gives clues about the origin of the rock. For example, the slow cooling needed to produce a coarse-grained rock such as **granite** would be most likely to occur in a large, deeply buried intrusion. The small crystals of fine-grained rocks such as **basalt** would suggest rapid cooling of smaller amounts of magma at or near the earth's surface.

Some crystalline igneous rocks do not fit neatly into one of the three grain-size categories because they contain large mineral crystals surrounded by other smaller crystals. This feature suggests that the magma cooled at two different rates. Such rocks are said to have a **porphyritic texture** and the individual large crystals are called **phenocrysts** (see Fig. 3.2 on page 33). The term **groundmass** is sometimes used to describe a mass of particularly small crystals surrounding phenocrysts.

Occasionally, small eruptions of magma may solidify so rapidly that there is no chance for any proper crystals to form. The rock type **obsidian** is produced in this way and shows a **glassy texture**. Obsidian is easily recognised by its shiny black colour and **conchoidal** (shell-like) fracture pattern. **Pumice** is also formed from rapidly solidified magma (volcanic froth) and shows no evidence of crystal growth.

Although pyroclastic rocks originate from magma they have different textures from other igneous rocks. They may contain mixtures of both crystalline and non-crystalline (glassy) fragments. Because they have 'settled' onto the earth's surface, pyroclastic rocks such as tuff and agglomerate often show sedimentary features such as **bedding**.

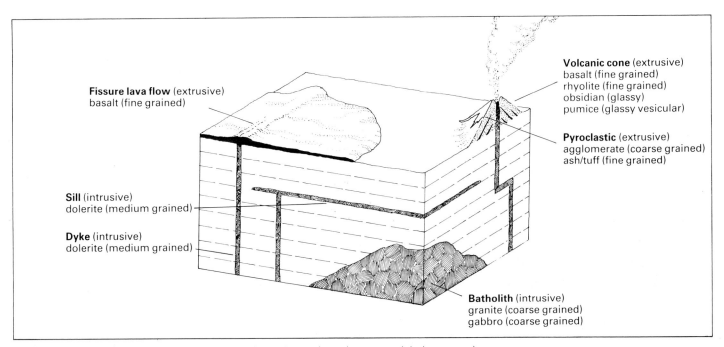

Fig. 5.16 Revision diagram for igneous activity (showing main rock types and their textures)

Mineralogy of igneous rocks

All igneous rocks are made up of silicate minerals. In crystalline specimens of coarse or medium grain size you should be able to recognise some of the following minerals.

- **Quartz**: identified as greyish 'glassy looking' grains, harder than steel.
- **Feldspars**: present in all igneous rocks: **orthoclase** has pink to white crystals while **plagioclase** is greyish white; both types are harder than steel and show faint cleavage planes.
- **Micas**: **biotite** is shiny black and **muscovite** has a silvery glitter; both micas cleave easily into thin flakes.
- **Mafic minerals**: this is the group name for dark-coloured (green to black) minerals such as **hornblende, augite** and **olivine**; since they all look very similar and have roughly the same hardness (6 on Mohs' scale) there is usually no need to distinguish between them.

When you discover which minerals are present in a rock, keep a record of their proportions. For example, you could end up with a list like this: specimen X contains approx. 25% quartz, 65% feldspar (mainly orthoclase) and 10% biotite mica.

Without a microscope it is impossible to study the tiny mineral crystals of fine-grained igneous rocks.

However, such rocks can be identified from their overall appearance. **Rhyolite** contains a lot of light-coloured low-density minerals (especially quartz and feldspar) so the whole rock is fairly light weight and grey to buff in colour. **Basalt** is full of darker denser minerals (especially the mafics) and is therefore black and heavy.

Naming igneous rocks

Apart from the pyroclastics which are identified by their 'sedimentary' features, all other igneous rocks are named according to their grain size and the minerals they contain. Fig. 5.19 is a chart which you can use to help identify igneous specimens. For example, if after studying a specimen you have decided that it is medium grained and contains mainly plagioclase feldspar and mafic minerals, then a look at the chart will tell you that the rock is dolerite.

It is worth noting that the main differences in the mineralogy of igneous rocks are related to the amount of silicon and oxygen they contain. The granite–rhyolite group has a high proportion of these elements (this explains the presence of the mineral quartz) and such rocks are often described as being 'acid'. The gabbro–basalt group have lower amounts of silicon and oxygen (hence no quartz is present) and can be termed 'basic'.

Fig. 5.19 Identification chart for common igneous rocks (the most important types are shown in bold letters)

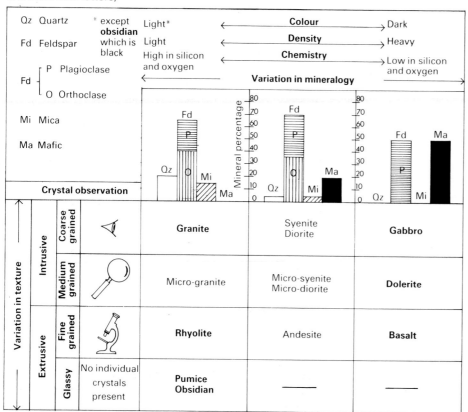

Typical description of an igneous specimen

Use this style in your practical work: the description refers to the specimen in Fig 5.18 on page 34.

Texture
- It is difficult to break individual minerals away from the specimen because they interlock together. Therefore this rock has a crystalline texture.
- Individual minerals (up to 5 mm in diameter) can be clearly seen without using a hand lens. Therefore this rock is coarse grained.
- The different minerals are evenly distributed throughout the rock and all have a similar size. The minerals do not appear to be aligned in any way.

Mineralogy
- Two different types of minerals can be seen in the rock.
- The greyish-white minerals are harder than steel and (under a hand lens) show faint cleavage planes. Therefore these are crystals of plagioclase feldspar.
- The dark green minerals are also harder than steel and show faint cleavage planes. They have a glassy lustre. Therefore these are crystals of mafic minerals (probably augite).
- The proportion of minerals is approximately 50% plagioclase and 50% mafics.

Other observations
- The weathered surface of the specimen shows a brown coloration suggesting that the rock contains a fairly high proportion of iron.
- The density of the specimen was calculated to be 3.0 g/cm^3.

Conclusion
The crystalline texture and random orientation of the minerals show that this is an **igneous rock**. Its grain size and mineralogy lead to the conclusion that it is a **gabbro**.

Questions

For questions 1–5 write the letters (A, B, C, ...) of the correct answers in your notebook.

1 Which **one** of the following factors provides the best evidence that an igneous rock was once molten?
A the presence of quartz
B the presence of mica
C the presence of mafic minerals
D the presence of ash particles
E the presence of a crystalline texture

2 Which **one** of the following rocks could be described as pyroclastic?
A tuff   B dolerite   C syenite
D basalt   E micro-granite

3 Which **one** of the following minerals would normally be present in both granite and gabbro?
A pyrite   B biotite mica   C plagioclase
D augite   E quartz   feldspar

4 Which **two** of the following rocks contain the same minerals as basalt?
A slate   B dolerite   C rhyolite
D gabbro   E pumice

5 Which **two** of the following rock types could not be formed by magma crystallising in an intrusion?
A agglomerate   B andesite   C syenite
D obsidian   E basalt

6 The following diagrams show details of three igneous rocks.

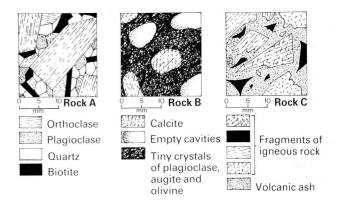

Look at rock A.
(a) What type of texture is shown?
(b) What is the correct name for the large orthoclase crystals?
(c) Name the rock type.
(d) What would be the two most abundant elements in this rock?
(e) Under what conditions do you think this rock cooled?

Look at rock B.
(f) What are the rounded empty cavities called and how were they formed?
(g) What are the calcite filled cavities called and where do you think this calcite could have come from?
(h) Name the rock type.
(i) What name could be given to the dark minerals olivine and augite?

Look at rock C.
(j) Name the rock type and explain how it could have been formed.
(k) What types of volcano tend to produce rocks like this?
(l) Do you think this rock was formed near to a volcanic vent or at a distance from one? Explain your answer.

6 Sediments and surface processes

Before beginning this chapter re-read page 27 to remind yourself of the main facts about sedimentary rocks. Remember that sediments are formed from 'second-hand' material which has been released from other older rocks by the actions of **weathering** and **erosion**.

Weathering

Weathering causes the gradual breakdown of rocks exposed to the earth's atmosphere. It is a process which involves the actions and interactions of many things (for example, rain water, temperature, atmospheric gases and bacteria). There are three main ways in which weathering occurs.

- **Physical weathering** occurs where physical forces cause the rock to disintegrate into smaller pieces. It happens mainly as a result of changes in temperature, especially in deserts, high mountains or polar regions.
- **Chemical weathering** occurs where chemical reactions cause the minerals in the rock to alter and decompose. It happens mainly in warm wet climates (warmth increases the rate of chemical reaction, and water allows chemical substances to be carried in solution).
- **Biological weathering** occurs where plants and animals cause the break-up of rocks. Fig. 6.2

Fig. 6.2 Growing tree roots can cause the break-up of rocks

shows how the growth of tree roots can force rocks to split. Burrowing animals can also help break up rocks.

You should not think that these three types of weathering act separately. For example, as physical break-up takes place it separates minerals from one another and allows chemical solutions to seep deeper into the rock. Similarly, chemical reactions may be helped by the biological actions of certain bacteria. In effect, all types of weathering make a combined attack.

Fig. 6.1 Weathering

Physical weathering

Frost shattering

Water expands when it freezes. To prove this, fill a small bottle right to the top. Screw on the cap and place the bottle inside a thick plastic bag before putting it into a freezer (the neck of the bag should be tied). After a few hours the expansion of the frozen water will have broken the bottle. Do not open the bag or handle the glass fragments.

When water enters a crack in rock and then freezes, it will force the crack open further. When the ice melts to water again, it can seep slightly deeper into the rock, so the next time it freezes it forces the crack open a little more. Repeated freeze–thaw action like this causes pieces of rock to break off. This often happens at the top of high mountains. The broken pieces of rock fall to the foot of the slope (see Fig. 6.3) where they lie as piles of jagged **scree**.

Fig. 6.3 Screes in Wasdale, Cumbria

Another experiment with frost action involves taking two sets of identical rock specimens, for example each containing a sample of shale, granite and sandstone. Keep one set dry but soak the other with water. Then place both sets in a freezer. Leave the dry set to freeze for several weeks but take the wet set out every day and allow the rock specimens to thaw before putting them back in the freezer (note: the specimens in this set must not be allowed to dry out). At the end of the experiment compare the specimens from each set. What can you conclude from the results?

Exfoliation

Minerals expand when they are heated and contract when they are cooled. If this happens repeatedly, the stresses can break a rock. The action is strongest in desert areas, where days are very hot but nights are cold. It is the outside of the rock which suffers most from heating and cooling so eventually thin sheets peel away leaving an **'onion skin'** appearance (see Fig. 6.4 on page 34). The correct name for this is **exfoliation**.

You can experiment with expansion and contraction by heating small pieces of rock in a bunsen flame and then plunging them into a sink of cold water. (Safety glasses must be worn.) Do not expect quick results because repeated expansion and contraction is needed. You could compare the effect on a crystalline rock (such as granite) with that on a sediment (such as sandstone).

Chemical weathering

This is mainly caused by the chemical actions of water and substances dissolved in it.

Some rock-forming minerals react with plain water. This is particularly true of feldspars, biotite mica, hornblende, augite and olivine. The reactions produce tiny mica-like particles known as **clay minerals**. These are easily transported away to be deposited eventually as a mud or clay sediment. The reactions that form clay minerals also release elements such as calcium, sodium, potassium, magnesium and silicon which dissolve in the water and are carried away.

Rain water often contains dissolved 'impurities' which increase chemical decomposition of rock. A typical example is shown by the reaction of rain water on limestone:

- rain water picks up carbon dioxide and becomes slightly acidic
  $H_2O + CO_2 \rightarrow H_2CO_3$ (carbonic acid)
- the acid reacts with calcite in the limestone
  $2CaCO_3 + H_2CO_3 \rightarrow 2Ca(HCO_3)_2$ (calcium bicarbonate)
- the calcium bicarbonate will dissolve and is removed in solution.

Fig. 6.5 Badly weathered stone pillar, York Minister

Fig. 6.6 How the weathering of granite supplies the 'sedimentary system'.
Draw your own diagrams to show the weathering of limestone and sandstone

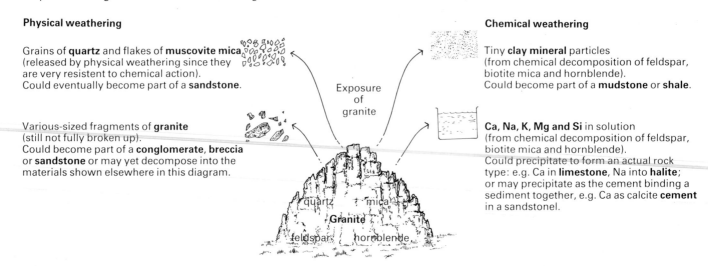

Physical weathering

Grains of **quartz** and flakes of **muscovite mica** (released by physical weathering since they are very resistent to chemical action). Could eventually become part of a **sandstone**.

Various-sized fragments of **granite** (still not fully broken up). Could become part of a **conglomerate, breccia** or **sandstone** or may yet decompose into the materials shown elsewhere in this diagram.

Exposure of granite

quartz   mica
Granite
feldspar   hornblende

Chemical weathering

Tiny **clay mineral** particles (from chemical decomposition of feldspar, biotite mica and hornblende). Could become part of a **mudstone** or **shale**.

Ca, Na, K, Mg and Si in solution (from chemical decomposition of feldspar, biotite mica and hornblende). Could precipitate to form an actual rock type: e.g. Ca in **limestone**, Na into **halite**; or may precipitate as the cement binding a sediment together, e.g. Ca as calcite **cement** in a sandstonel.

Chemical weathering often alters the colour and appearance of rocks. When visiting field locations always choose a fresh unweathered specimen (hammer only if necessary) to see what the rock is really like. You will notice that sometimes weathering occurs further in than just at the exposed surface.

Some weathered rocks show a rusty red/brown surface stain. This is due to insoluble iron compounds left by the chemical reactions. The deep red colour of tropical soils is caused by the same thing.

Although weathering is a process of rock destruction, you should remember that in a sense it is also responsible for rock creation. The weathered materials will travel on to become new sediments, and eventually new rock. (See the example given in Fig. 6.6.)

Transportation, erosion and deposition

Weathering produces rock, mineral and chemical **debris** at the place where the original rock was exposed ('in situ'). This material is usually moved or **transported** to a different site before being deposited as a **sediment**. The material can be moved in several different ways:
- by **gravity**;
- by water in **rivers** and **streams**;
- by water in **the sea**;
- by **glacial ice**;
- by **wind**.

Often several of these 'agents' of transportation may move one sediment. For example, pieces of rock may fall (gravity) onto a glacier; then be carried by ice until melting occurs; then be moved on by a stream; then reach the sea; and eventually be deposited by the waves.

The agents of transportation not only move sediment; they also produce more of it by wearing away the rocks they travel over. This process is called **erosion**. It involves such things as glacial ice cutting a valley or sea water breaking down cliffs.

Fig. 6.8 Transportation and rounding

Freshly weathered (e.g. scree)   Partly rounded (e.g. on stream bed)   Well rounded (e.g. on beach)

Fig. 6.7 Transportation: the 'conveyor' of sediment

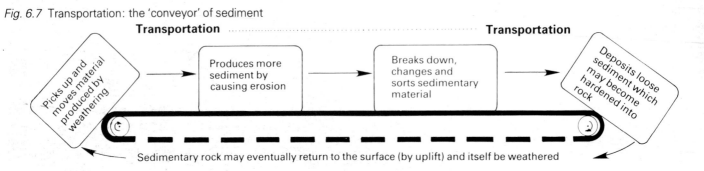

Transportation

Transportation

Picks up and moves material produced by weathering

Produces more sediment by causing erosion

Breaks down, changes and sorts sedimentary material

Deposits loose sediment which may become hardened into rock

Sedimentary rock may eventually return to the surface (by uplift) and itself be weathered

The main power of erosion comes from the material being carried. For example, the pebbles carried in a stream cause more damage to the solid **bed rock** beneath than the water itself. The action of the load as a cutting tool also affects the load fragments themselves. As they strike each other and the surrounding rock they too become damaged. Fragments become smaller as they travel on, and may become rounded as the corners are knocked off (see Fig. 6.8).

The agents of transportation can only carry sediment if they have the necessary energy. When energy is reduced, some of the load must be deposited. For example, rivers deposit sediment when they flow into a lake, because the water loses energy. Some forms of deposition do not depend on the energy of transportation, however. Material can be brought out of solution by certain chemical reactions (for example, halite is **precipitated** from evaporating sea water) or the action of living things (for example, animals extract calcium carbonate from water to build their shells, and a deposit of shells can make a rock). We shall study these processes of deposition in more detail in Chapter 7.

The different styles of transport and the different reasons for deposition act as a **sorting** system for sedimentary material. Although a wide variety of sediment will be produced as an area is weathered and eroded, this will not all be deposited in one place. Larger fragments will tend to settle together, while the finer material will be carried on to other sites. The chemical load will only be deposited where conditions are suitable for it to come out of solution. The overall

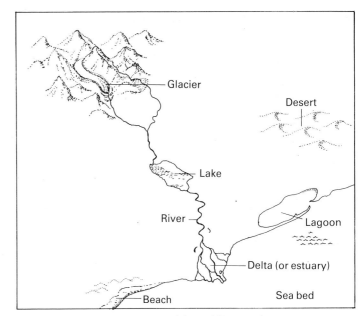

Fig. 6.10 Main environments of deposition

effect is to sort out the mixture of sediment into different types. Each type of sediment will in turn produce a different rock type in the particular environment in which deposition took place (see Fig. 6.10).

As erosion and deposition take place, the landscape is changed. The removal of rock material produces **erosional features** such as valleys, caves and cliffs. The laying down of sediment produces **depositional features** such as beaches, dunes and deltas. We shall now study the erosional and depositional effects of each of the main types of transportation.

Fig. 6.9 Deposition of sand and gravel in a river channel. *Do you think this is a permanent deposit? What is likely to happen when the river is in flood?*

The effect of gravity on sedimentary material

Weathered debris is often loose and unstable so it tends to move downhill under the influence of gravity. We have already seen how this produces **scree slopes** (Fig. 6.3). On more gentle slopes, movements may be a slow but steady **soil creep** which produces the features shown in Fig. 6.11. On steeper slopes such as cliffs, valley sides or even road cuttings, sudden **rockfalls** or **landslips** may occur. Most sudden slips need a 'trigger action' such as heavy rain or a rapid spring thaw to set them off. They are also more likely to happen where the underlying bed rock is arranged in an unstable formation (see Fig. 6.12a, b).

Gravitational movement of sediment is one of the most important processes in the **denudation** (wearing away) of the land. Without it, weathered material would not 'feed' downhill into streams, etc. Gravity is also responsible for the movement of rivers and glaciers and the final settling of all sediments during deposition.

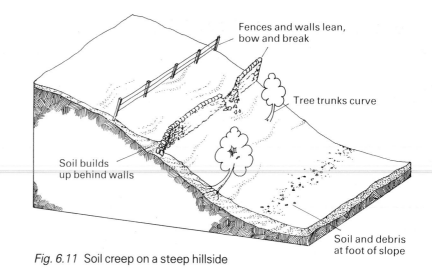

Fences and walls lean, bow and break

Tree trunks curve

Soil builds up behind walls

Soil and debris at foot of slope

Fig. 6.11 Soil creep on a steep hillside

Fig. 6.12a Landslip at Mam Tor, Derbyshire

Fig. 6.12b Collapsed and abandoned road below Mam Tor

46

The geological effect of running water

Rivers and streams act as part of the return system which drains rain water back into the oceans. In doing so they transport, produce and deposit sediment.

Transportation by running water

Running water carries sediment in three ways.

Bed load. This is the larger material such as pebbles and sand grains which is pulled along with the water. Considerable energy is needed to move it but once in motion it adds 'cutting power' to the flow and causes erosion. The bed load moves by sliding, rolling and bouncing along, but this does not happen constantly. Often the fragments are 'dumped' for a while then moved again when the flow of water increases after heavy rain.

Suspension load. This is the fine silt and mud (clay minerals) which is carried suspended within the water. These smaller particles require less energy to carry than the bed load and are only deposited when the flow is very low in energy. In some cases, especially during floods, river water has so much suspended sediment that it looks muddy and brown.

Solution load. This is the invisible load of dissolved chemicals which have been released during weathering. In some rivers, industrial pollution may add to this material.

River erosion

The following factors combine to cause erosion along the bed and sides of a river channel. They also cause fragments of sediment carried by the river to become smaller and/or more rounded.

- the power of the moving water itself,
- the impacts of the bed load and suspension load,
- the chemical reactions between the water and rock.

Fig. 6.13 Formation of a waterfall (*Note*: the layers of rock do not have to be horizontal to produce this effect)

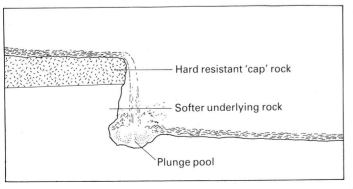

Hard resistant 'cap' rock

Softer underlying rock

Plunge pool

Fig. 6.14 Development of valley shape

(a) Normal 'V'-shaped valley

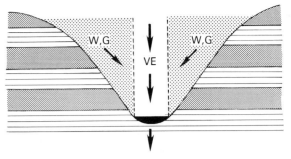

(b) Steep-sided gorge

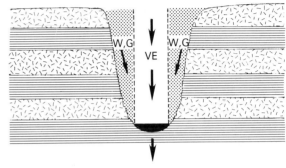

VE   Zone of rock removed by vertical erosion as river cuts along its bed

W,G   Zone of rock released by weathering and carried by gravity into river channel

Erosion occurs at different rates as the river flows over different rock types. This may cause a waterfall (Fig. 6.13) to be formed, for example. Other geological factors can affect the rate of erosion. For example, a river may erode more easily along the weakened rocks of a fault line.

The shape of a river valley can also vary. Normally a river cuts a channel by eroding the rocks immediately beneath it. As the cutting action continues, the channel becomes lower and lower. The steep valley sides produced by this **vertical erosion** are gradually worn back by weathering, and gravity causes the weathered material to fall into the river. In this way, a more gentle 'V'-shaped valley develops. As Fig. 6.14a shows, a valley tends to become wider as the river cuts deeper. In exceptional circumstances (such as where the rock is very hard or where a dry atmosphere restricts chemical weathering), the valley sides will remain steep. In such cases a gorge (Fig. 6.14b) is formed.

The ability of a river to deepen its valley depends partly on sea level. If the sea level is lowered (or the land is raised), river channels can then cut down to a lower base level and the vertical erosion increases.

River deposition

The energy of a river depends on the amount of water flowing and its speed of travel. If energy is reduced, some of the river's load must be deposited. This can happen at various places in a river's course, producing different types of deposition.

River bends

Fig. 6.15 shows that sands and gravels are laid down on the inside curve of bends. Notice too that the bank on the outer curve is **undercut** by erosion. Over long periods of time this can widen the valley area.

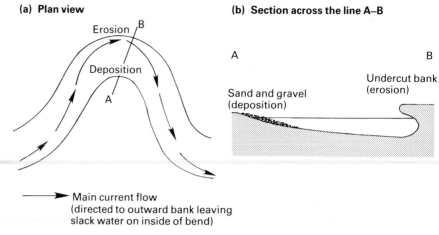

(a) Plan view

Erosion

B

Deposition

A

→ Main current flow
(directed to outward bank leaving slack water on inside of bend)

(b) Section across the line A–B

A

B

Sand and gravel (deposition)

Undercut bank (erosion)

Fig. 6.15 Erosion and deposition on a river bend

Fig. 6.16 Section across a river flood plain

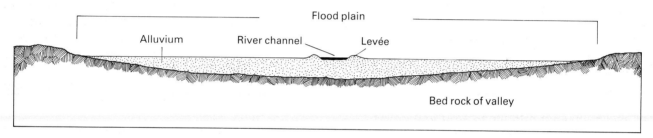

Flood plain

Alluvium   River channel   Levée

Bed rock of valley

Flood plains

These are broad flat areas bordering large lowland rivers. During a flood, water flows from the channel and loses some of its energy by spreading over a wider area. As it does so, it deposits some of its load. Coarse sand and gravel are deposited first, just beyond the original bank sides. After repeated 'flood dumping' this area becomes built up into **levées** (raised river banks: see Fig. 6.16). Fine silt and mud is carried further out and deposited right across the flood plain as a bed of **alluvium**. Some of the world's best soils are on the alluvial flood plains of great rivers like the Ganges and Nile.

Alluvial fans

These deposits occur where powerful streams lose energy by flowing out onto flatter land. Fig. 6.17 shows a typical example where a fan has formed at the foot of a line of steep hills. Alluvial fans are deposited rapidly, so a mixture of poorly sorted sand and pebbles is found. By studying the position of alluvial fans formed in the past, geologists can work out where ancient mountain ranges once met lowland regions.

Fig. 6.17 Alluvial fan

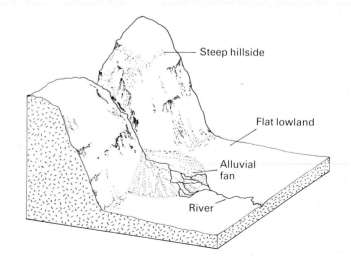

Steep hillside

Flat lowland

Alluvial fan

River

Deltas

When a river flows into a large body of water such as the sea, energy is quickly lost and deposition occurs. Fig. 6.18 shows a typical delta shape with each **distributary** 'distributing' its share of water and sediment. Not all river mouths produce deltas. Often wave action is powerful enough to keep the sediment moving and so it is deposited elsewhere. Depth of water and rate of sediment supply also affect the style of delta growth.

Large rivers can build massive deltas. For example, the Mississippi is thought to deposit nearly two million tonnes of material each day. Over long periods of time, thick and wide expanses of delta sediments create special sequences of sedimentary rocks. Later in this book you will learn how Britain's coal seams were formed (about 300 million years ago) as part of a deltaic sequence.

Lakes

A lake provides a stretch of calm water along a river course. As the running water loses energy in the lake, part of the river's load is deposited, so the lake becomes a sediment trap. In geological terms, lakes are only temporary features because they slowly become infilled with sediment. (This happens in reservoirs too, often creating problems.)

Pockets of alluvium often indicate the position of an ancient lake which became infilled with sediment many years ago. In more recent examples the surface of such areas will still be marshy. Some past lakes have disappeared not by infilling but by evaporation.

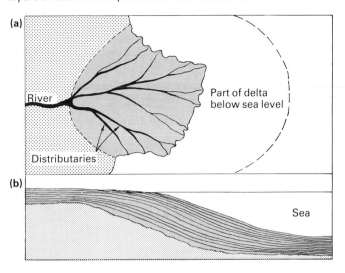

Fig. 6.18 River delta: (a) plan view; (b) section showing sloping layers of sediment deposited as delta builds out

An interesting project is to make a list of all the geological situations which can produce lakes. These will include places where a sunken area has been caused (allowing a body of water to collect) or where a river valley has been dammed by some natural event. You will need to read more before you can make a full list but some examples from igneous activity might include:

1. *a hollow made by the vent of an extinct volcano (example: Crater Lake, U.S.A.);*
2. *the damming of a valley by a flow of lava (example: the Sea of Galilee, Israel).*

Fig. 6.19 Buttermere and Crummock Water (Lake District): two lakes partly infilled and now separated by an alluvial fan

Underground water

Water can pass through certain types of rock. There are two ways in which this happens.

1. **Porous rocks** have tiny spaces between their mineral grains which allow water to seep in. The most porous rock type is **sandstone** (see Fig. 6.20).
2. **Permeable rocks** have a series of gaps and fractures within them which act as a natural 'plumbing system' for water to pass through. For example, water can often seep along the junctions between beds of sedimentary rock (**bedding planes**), or move along **joints** (almost vertical cracks caused by earth forces: see page 101). One of the most permeable rock types is **limestone** (see Fig. 6.21). **Impermeable** rocks do not allow any water to pass through.

Some **groundwater** (water in rock below the surface) finds its way naturally back to the surface at a **spring**. In other cases it provides an underground 'reservoir' to be drawn up from **wells** (Fig. 6.22). The geology of water supply is also dealt with on page 186.

Groundwater presents problems in some mines and may require special drainage tunnels and/or powerful pumps to remove it. Many old mines are now in a very dangerous flooded state. You should never enter such places without an expert guide.

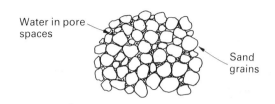

Fig. 6.20 Close-up view of porous sandsone. The porosity (the amount of water that can be contained in the rock) depends on how much compaction and cementation have affected the sediment

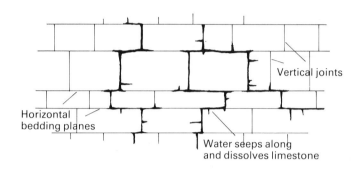

Fig. 6.21 Close-up view of permeable limestone

Fig. 6.22 Spring and wells

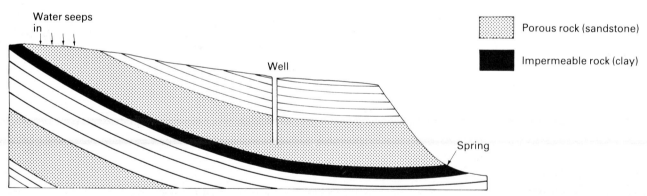

Fig. 6.23 Underground water being used to irrigate desert lands near the Tunisia/Algeria border

Fig. 6.39 Glacial till forms the easily eroded cliffs of Holderness (South Yorkshire). The small boat is being lowered into the water because there is no suitable harbour

Fig. 6.44 Seif dunes in the Namib Desert, south west Africa

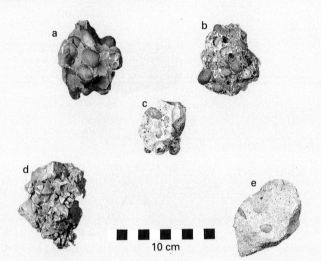

Fig. 7.5 Specimens of conglomerate (a, b, c), breccia (d) and pebbly sandstone (e)

Fig. 7.9 Specimens of sandstone (a, b, c), clay (d), siltstone (e), mudstone (f) and shale (g, h)

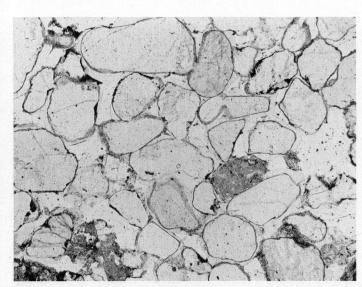

Fig. 7.10 Photomicrograph of desert sandstone. Note the fragmental texture of quartz grains coated with a (red) iron oxide cement. The area in view is about 1.5 mm wide

Fig. 7.13 Specimens of chalk (a), muddy limestone (b), shelly limestone (c) and oolitic limestone (d)

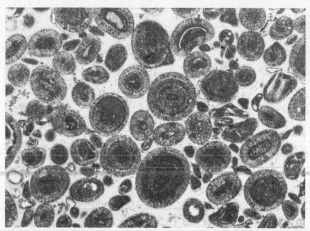

Fig. 7.15 Photomicrograph of oolitic limestone. The area in view is about 5 mm wide

(a)

Metamorphism →

10 cm

(b)

Metamorphism →

Fig. 8.6 Specimens showing difference between (a) sandstone and metaquartzite, (b) limestone and marble

Fig. 7.33 Nodules (N) in shale

Fig. 7.34 Specimen of sandstone: see page 78 for full description

a

b

Fig. 8.8 Specimens showing difference between mudstone (a), slate (b), schist (c) and gneiss (d)

c

d

Increasing grade of metamorphism

10 cm

5 cm

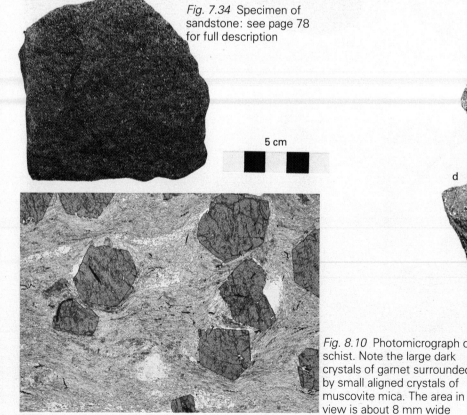

Fig. 8.10 Photomicrograph of schist. Note the large dark crystals of garnet surrounded by small aligned crystals of muscovite mica. The area in view is about 8 mm wide

Underground drainage in limestone areas

You have already learned that rain water causes weathering of limestone (page 43). The same effect occurs below ground where water reacts with and dissolves limestone as it seeps along bedding planes and down joints. In time, these small openings are widened into larger passages which can carry underground rivers and streams.

As solution of the limestone continues, water often breaks through the rock to open new routes along the maze of cracks. Passages which were once full of moving water may be left dry or with just a small stream flowing through them. These drier parts of the system become sites of chemical precipitation. Drops of water leave tiny amounts of calcium carbonate ($CaCO_3$) as they drip from the cave roof or trickle down the side walls. As precipitation continues, **stalactites** are produced on the cave roof and **stalagmites** on the cave floor.

Fig. 6.25 shows the typical features of underground drainage in a limestone area. Notice the vertical **sink hole** where the stream disappears (Fig. 6.26) and the **spring** (in this case a cave mouth) where it returns to the surface. Landscapes like this are a feature of many parts of the Pennines, especially the Yorkshire Dales and the Derbyshire Peak District.

Junctions between limestone and other rocks can often be mapped by studying the positions at which streams disappear and reappear.

Fig. 6.24 Stalactites and stalagmites in East Gill Caverns (northern Pennines)

Fig. 6.26 Disappearing stream, Malham, Yorkshire

Fig. 6.25 Drainage in a limestone region

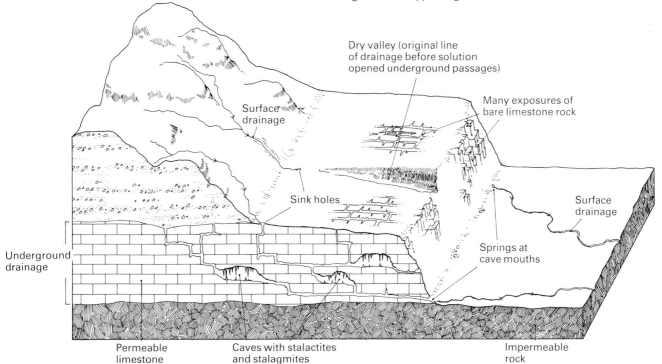

Dry valley (original line of drainage before solution opened underground passages)

Many exposures of bare limestone rock

Surface drainage

Sink holes

Surface drainage

Underground drainage

Springs at cave mouths

Permeable limestone

Caves with stalactites and stalagmites

Impermeable rock

The geological effect of the sea

The geological work of the sea is easiest to study at the **coast**. Here waves, currents and tides produce distinct features of coastal erosion, transportation and deposition. However, the sea is not always the only factor involved in shaping a coast. For example, river action will be partly responsible for the shape of deltas and estuaries, while onshore winds (blowing from the sea) may produce coastal sand dunes.

Waves

Waves are produced by wind blowing over a surface of water. The size of waves depends on wind speed, water depth and the distance of water over which the wind has blown. The largest waves are formed by galeforce winds blowing across wide deep oceans.

Waves at sea do not carry water forward with them. The water moves rhythmically up and down giving a pattern of alternate crests and troughs. In shallower water, close to shore, this movement is affected by the sea bed. Because the trough cannot develop properly the crest is forced to rise up until it topples forward and **breaks**. At this point water moves up onto the shore in what is called the **swash** of the wave. When the energy of this is spent, water will return downslope as the **backwash**. As Fig. 6.27 shows, the power of the swash and backwash depends on the size of the wave and the slope of the shore. You have no doubt felt the powerful backwash (sometimes called **undertow**) of waves breaking on a steep beach.

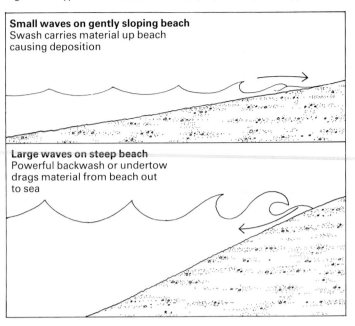

Fig. 6.27 Types of waves

Small waves on gently sloping beach
Swash carries material up beach causing deposition

Large waves on steep beach
Powerful backwash or undertow drags material from beach out to sea

Coastal erosion and transportation

The most intense erosion occurs on coasts which face large oceans and are often attacked by storm waves. The force of rushing water, and the impact of rock fragments thrown against the shore, are powerful aids to erosion. The chemical action of salty water also causes break down, as does the rapid increase in pressure caused by crashing waves trapping and compressing air in rock crevices.

Fig. 6.28 Cliff erosion

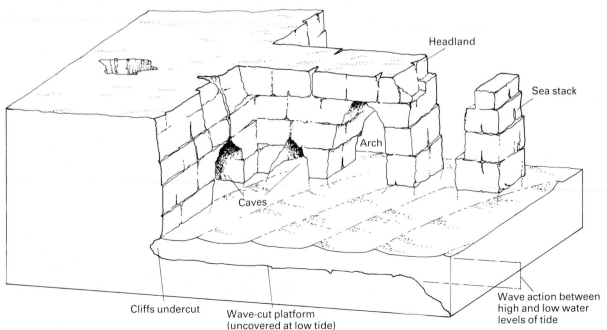

Headland

Sea stack

Arch

Caves

Cliffs undercut

Wave-cut platform (uncovered at low tide)

Wave action between high and low water levels of tide

54

The sea erodes (**undercuts**) the rock at the base of cliffs until whole sections fall away. In time, a **wave-cut platform** (see Fig. 6.28) is left as the cliff line is pushed back. The rate of undercutting will depend on the type of rock being attacked. Soft sedimentary rocks with weaknesses such as bedding planes and joints will 'retreat' faster than hard igneous or metamorphic rocks. Variations in rock type produce a coastline of headlands and bays (see Fig. 6.29) as the rock is eroded at different rates.

At weaker points in cliffs (for example, at the site of larger joints), **sea caves** may be gouged out by the waves. If two caves are cut back-to-back in a headland, an **arch** may be produced. If the top of this falls away, a **sea stack** is left (see Fig. 6.28).

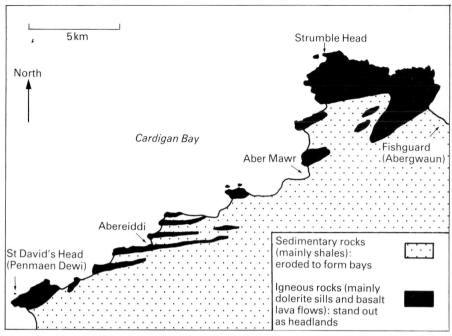

Fig. 6.29 Geology and coastal form: part of Dyfed coast, Wales, showing bays and headlands produced by different rates of marine erosion

As well as attacking shorelines, waves and currents also transport sediment. Much of this sediment will have been produced by coastal erosion but some may be from other sources: for example, material brought to the sea by rivers.

Sediment may be moved along the coast when waves meet the shore at an angle. Fig. 6.30 explains this process of **longshore drift**. In many places, however, the movement is stopped by a headland acting as a barrier. Some holiday resorts build their own barriers of wooden **groynes** (See Fig. 6.31) to prevent longshore drift from 'stealing' the beach.

Fig. 6.30 Longshore drift (plan view)

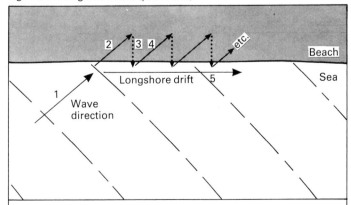

1. Waves approach the beach at an angle.
2. Pebbles and sand are washed up onto the beach in the direction of wave.
3. They roll straight down the slope with backwash of wave.
4. Next wave repeats the process.
5. Result: material 'drifts' along the shore.

Fig. 6.31 Groynes along the coast at Eastbourne, England. *The photograph was taken looking south west; in which direction is longshore drift acting?*

Marine deposition

Most of the world's sedimentary rocks were originally deposited on a sea bed. To understand these rocks properly geologists need to study what kind of sediment is forming beneath present-day oceans. Although 71% of our planet is covered by sea it is only recently that this huge area has been scientifically investigated. Survey ships with sonar and echo-sounding equipment have enabled the ocean floor to be mapped. Underwater photography and samples collected from the sea bed have given evidence of the sediments to be found there.

To study marine deposition it is convenient to divide the oceans into four depth zones (see Fig. 6.32).

Inter-tidal zone

This is the shoreline where wave action provides the energy for erosion of the coastline and deposition of beach sediment. **Beaches** develop where the backwash is not powerful enough to drag sediments (which have been deposited by the swash) back out to sea.

Beach sediments of sand and pebbles tend to become well rounded because each particle is constantly washed back and forth by the breaking waves. The continuous rolling and rubbing together is also likely to break down weaker rock and mineral fragments completely. Only hard resistant materials survive in this environment. For this reason beach sands are largely composed of resistant quartz grains, and beach pebbles are mostly of hard igneous and metamorphic rock types. If beach sands become

Fig. 6.33 Sand spit at Spurn Head, formed by the action of longshore drift across the mouth of the Humber Estuary

buried and compacted, they usually form sandstones known as **orthoquartzite** (page 68). Beach pebbles become **conglomerate** (page 66).

Apart from beaches, other sediment deposits in the intertidal zone include **sand banks** and **spits** (Fig. 6.33) produced by current and/or longshore drift movement. Finer sediments of mud and silt (mud flats) may also be deposited in the quieter conditions of estuaries. Evaporation from coastal lagoons in dry regions can give **evaporite deposits** (see page 72). Beds of **coals** may even form in the swampy conditions of coastal deltas (see page 71).

Fig. 6.32 Section showing the zones of the ocean floor: (a) inter-tidal zone, (b) continental shelf, (c) continental slope, (d) ocean trench, (e) abyssal plain. Note the difference between the vertical and horizontal scales. This greatly exaggerates the angles of slopes

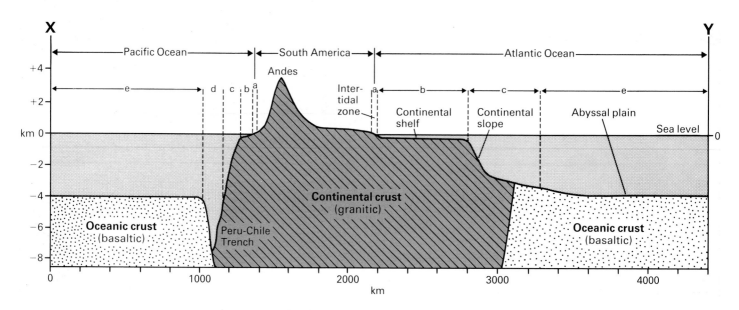

Continental shelf

Continental crust extends beyond the shore to form a shelf (see Fig. 6.32). The sea above this is shallow, usually less than 250 metres deep. As Fig. 6.34 shows, continental shelves vary considerably in width. For example, the eastern (Atlantic) shelf of South America is up to 1000 km wide but its western (Pacific) shelf only averages 100 km. The British Isles are actually blocks of higher ground rising from the wide Atlantic shelf of Europe. The whole of the North Sea lies on top of continental crust.

Continental shelves provide a great variety of sedimentary environments and the deposits being formed will depend on such factors as:

- the amount and type of sediment being supplied from the nearby continent;
- the width and shape of the shelf area;
- the direction and strength of sea currents;
- the temperature and chemical composition of the sea water.

Despite so much variation, some general points can be made. Sands are the main deposits in areas of current action (see page 68 for details of the **sandstones** which are produced). Fine silts and muds tend to build up in quieter offshore waters, but may be found nearer to the coast where there is a suitable supply from a river mouth (see page 69 for **siltstones, mudstones** and **shales**). **Limestones** tend to be deposited on shelves where the water is clear, warm and shallow but only a very small amount of sand or mud is being supplied (see page 70).

Deposition on continental shelves is a vital part of the rock cycle (page 28). At some stage, sedimentary rocks formed here are usually uplifted and 'returned' to become part of the land once more. The fact that much of the surface of the continents is made of marine sedimentary rocks shows that (at certain past times) they have been flooded as shelf areas.

Continental slope

The continental slope begins at the edge of the shelf (on average about 150 metres deep at this point) and descends to the deep ocean beyond (several kilometres below sea level). Although its average angle is only 4°, it is considered to be a steep slope in the ocean where gradients are generally much more gentle than those we are used to seeing on land.

The type of sediment that reaches the continental slope depends on the width of the shelf and the currents acting on it. On narrow shelves, sand and silt may be carried out to the slope. But on wider shelves it is likely that only finer clays will travel so far from land.

Continental slopes are cut by huge **submarine canyons** (often several hundreds of kilometres long and over a kilometre deep) which act as undersea 'rivers' carrying material towards the deep ocean. Within these canyons, and along the slope itself, sediments often become unstable and begin sliding or slumping towards the deeper areas. Such movements are often caused by shock waves from sea bed earthquakes. They result in clouds of sediment

Fig. 6.34 Continental shelves and ocean trenches

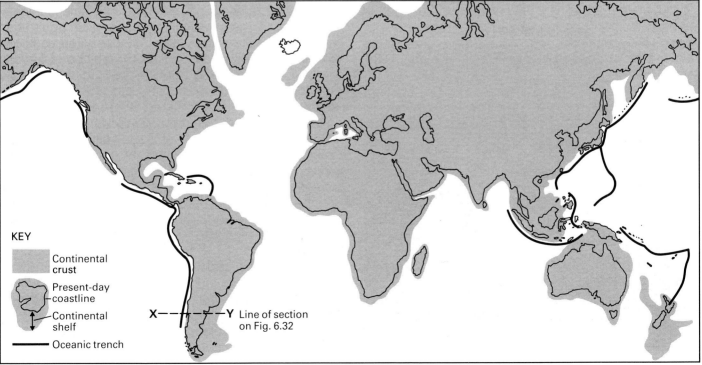

KEY

- Continental crust
- Present-day coastline
- Continental shelf
- Oceanic trench

X - - - Y Line of section on Fig. 6.32

being dispersed into the water. These clouds travel rapidly downslope (like undersea 'avalanches') and are called **turbidity currents**. As they lose energy at the base of the slope, turbidity currents deposit their load to give sediments such as **greywacke** (see pages 68 and 76).

Continental slopes such as the one off the west coast of South America (Fig. 6.32) descend 6 to 8 kilometres directly into ocean trenches. In this region, rapid erosion of the nearby Andes Mountains (plus the narrow width of the shelf) mean that parts of the Peru–Chile Trench are being infilled at the rapid rate of over a kilometre every million years.

Deep ocean floor

A large part of this zone consists of flat smooth areas (often 4 km below sea level) called **abyssal plains**. Along the edge of these plains (near the foot of the continental slope) some fine muds may be delivered by turbidity currents, but very little material is carried further or deeper than this. Across the rest of the plain only a thin 'veneer' of sediment exists. This is either **clay** which has been formed by wind-blown dust and volcanic ash falling and sinking far from land or **ooze** formed from the minute skeletons of microscopic sea creatures.

Fig. 6.36 Carrick Roads, a ria in south west England (the town of Falmouth can be seen in the foreground)

The level and extent of the sea

The position of a coastline depends on the level of the sea and the level of the land. Although these levels may appear to change little during a human lifetime, they alter slowly but with dramatic effect during geological time.

As time passes, sea level changes because (a) the volume of ocean basins changes and (b) the amount of water in these basins varies. Plate movement is responsible for altering the volume of ocean basins (this will be explained in more detail on page 148).

The amount of water in the oceans changes particularly when the earth's climate becomes colder and ice ages occur. At these times so much water is frozen (and therefore trapped) on land that there is less in the oceans. During the most recent ice age, about 1 million years ago, sea level was about 150 m lower than it is today.

The level of the land changes with time because earth forces cause uplift or subsidence. The largest movements take place where continental shelves are uplifted and deformed into mountain chains by the force of plates moving together (pages 16 and 144). Other uplifts and subsidence can be the result of **isostasy** (page 17).

The amount that sea and land levels have changed is well illustrated by the 'shale band' (which causes problems for climbers) just below the top of Mount Everest. Shale is a sedimentary rock, which shows that this high mountain region was originally part of a continental shelf covered in sea. Similar evidence of changing levels exists in Britain where, for example, marine sediments are now exposed in the Pennines and Welsh hills.

Features produced by recent changes in level can be seen along present-day coastlines. Fig. 6.35 shows a **raised beach** which has been left 'high and dry'. Fig. 6.36 illustrates the opposite effect where the lower parts of a river valley have been 'drowned' by a rise in sea level, producing a **ria**.

Fig. 6.35 Raised beach

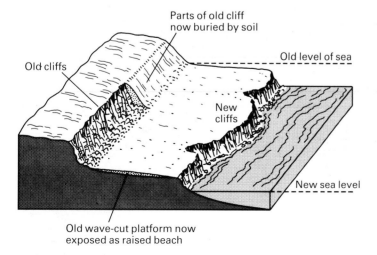

Parts of old cliff now buried by soil

Old level of sea

Old cliffs

New cliffs

New sea level

Old wave-cut platform now exposed as raised beach

The geological effect of ice

At present about 10% of the earth's surface is covered by ice. Much of this is simply frozen sea water (for example, the Arctic Ocean) which has fairly limited geological effects. In this part of the book we concentrate on the solid water which covers parts of the land surface.

Ice forms on land when the temperature is low enough for snow to exist all year round. As layer upon layer of snow builds up, the lower levels become compressed into a mass of solid ice. This ice will begin to move under the influence of gravity and a **glacier** develops.

Glacier ice covers the land in different ways. In mountain regions such as the Alps and Andes it fills whole valleys, carrying ice to lower warmer levels where it melts. In colder areas valleys of ice may join to spread out as a sheet (see Fig. 6.37), covering the lowland as well. In extreme cases whole land masses may be totally buried: for example, Greenland and Antarctica where the ice is up to 3 km thick.

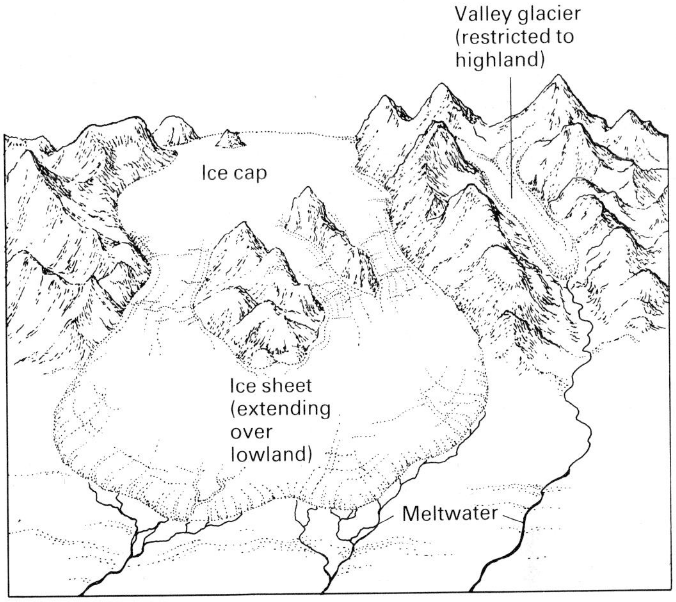

The movement of ice

Studying glacier movement is difficult because the process is very slow (usually less than 1 metre per day). The glacier remains solid overall but is able to flow as individual ice crystals slide against each other. Further movement occurs where thin films of water allow sliding to take place.

Fig. 6.37 Types of glacier

Often different parts of the glacier move at different rates causing huge cracks (**crevasses**) to open up. Crevasses are commonly found where the glacier pushes up and over an obstruction in the valley floor.

Fig. 6.38 Section along a valley glacier

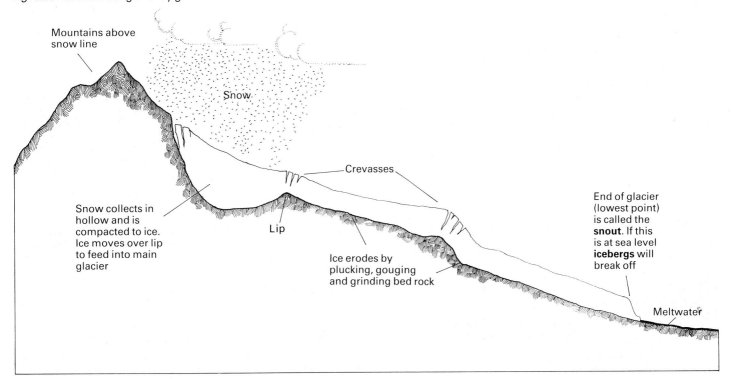

The power of ice: erosion and transportation

When glacier ice moves over a land surface it erodes by **plucking, gouging** and **grinding**. Plucking occurs when pieces of bed rock become frozen into the base of the ice and are then plucked out when the glacier moves slowly onwards. Debris collected by plucking provides the base of the ice with 'cutting tools' which cause further erosion by grinding and gouging the rocks beneath.

As ice-carried fragments grind against bed rock they become ground down into dust particles called **rock flour**. This material can act like fine abrasive powder smoothing the rock surfaces that it moves over. Gouging by sharper fragments may mark these smoothed surfaces with deep scratches called **glacial striations**. If you find examples of such striations during fieldwork trips try to work out which way the ice must have been moving to produce them.

Although most of a glacier's load is picked up by erosion of the bed rock beneath, some material may be collected and carried on its surface. This is frost-shattered debris that has fallen onto the ice from surrounding mountain peaks.

Estimates suggest that a valley glacier has twenty times the eroding power of a stream carrying the same amount of water. When you imagine the force of thousands of tonnes of ice 'armed' with boulders and rock fragments frozen to its base, it becomes easy to see how wide deep valleys can be formed or even whole regions worn down to bare smooth rock.

Glacial deposition

A mass of moving ice can transport huge amounts of sediment which vary in size from fine rock flour to enormous boulders. Because of this huge size variation, a key feature of glacial deposits is their mixed unsorted nature. When the ice melts it simply dumps all its load together.

Dumping of material at the snout of a glacier produces banks of **terminal moraine** (terminal means end). These moraines often form a series of ridges across glacial valleys. Each ridge marks the position of the snout at a particular stage during the final melting and retreat of the glacier. Terminal moraines may act as dams and cause lakes in some glacial valleys.

When a large ice sheet has extended across a wide area, the eventual melting will cover most of the ground surface with **till**. As Fig. 6.39 on page 51 shows, till is a mixture of glacial debris (from boulders to clay) all dumped together.

Ice sheets can transport rocks for large distances. Sometimes a particularly distinctive rock type can help geologists to discover the direction and distance of ice movement. For example, granite from the intrusion at Shap in the Lake District has a unique appearance because of its prominent pink phenocrysts of orthoclase feldspar (see Fig. 3.2 on page 33). The fact that boulders of Shap granite are found in till on the Yorkshire coast shows that ice sheets must have moved eastwards right across the northern Pennines. Distinctive rock types dumped far from home are called **glacial erratics**.

Fig. 6.40 Glaciated highland landscape, Haweswater, Lake District

Glacial meltwater

Melting releases large amounts of water which pick up some of the glacial debris (especially sand and clay-sized particles) and transport them elsewhere. Terminal moraines and till are often partly washed away by meltwater. Sediment is carried beyond the area of the original ice sheet and, as the water loses energy, the sediment is deposited to produce well-sorted beds of sands and clays.

Large temporary lakes form where ice or high ground prevents the escape of meltwater. Such lakes may leave deposits of sands and clays which show distinct layers marking the different rates of supply in winter and summer. Meltwater also has erosive power and can cut its own channels or deepen existing river valleys.

Before escaping from glaciers, meltwater may have flowed within them as **sub-glacial streams**. These form tunnels through the ice and usually deposit sediment as they flow. After the ice has melted the tunnel sediment can be seen as a winding sandy ridge called an **esker**.

Glaciation in Britain: the Ice Age

The Ice Age in Britain lasted from about 1 500 000 to 10 000 years ago and much of Britain's present landscape shows clear evidence of glacial erosion or deposition.

Fig. 6.41 gives before, during and after views of erosion by valley glaciers in highland areas such as North Wales and Scotland. The features shown include: glacial valleys (troughs) (example: Nant Ffrancon, Snowdonia); arêtes (example: Striding Edge, Lake District); and corries (example: those on the sides of Ben Nevis). Many glacial valleys have been flooded by lakes (example: Thirlmere, Lake District) or flooded by the sea (example: Loch Leven, Scotland).

Ice sheets spread across lowland Britain too, leaving moraines, till or meltwater sands and clays across all but the southernmost part of the country.

We shall consider the Ice Age again on page 170 as part of our study of Britain's geological history.

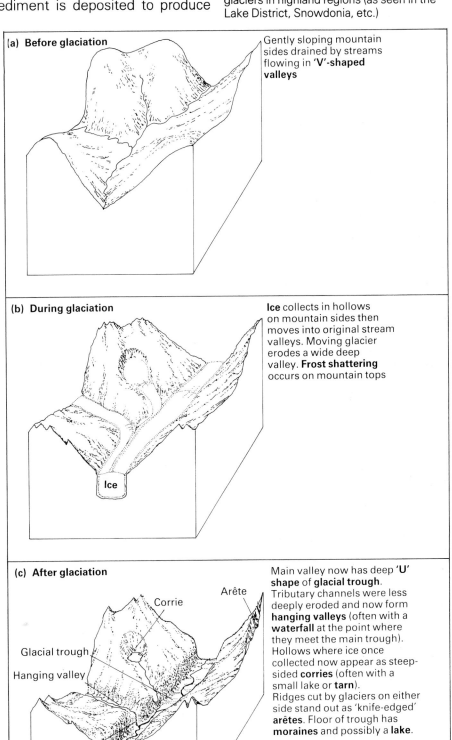

Fig. 6.41 Features produced by valley glaciers in highland regions (as seen in the Lake District, Snowdonia, etc.)

(a) Before glaciation

Gently sloping mountain sides drained by streams flowing in **'V'-shaped valleys**

(b) During glaciation

Ice collects in hollows on mountain sides then moves into original stream valleys. Moving glacier erodes a wide deep valley. **Frost shattering** occurs on mountain tops

Ice

(c) After glaciation

Arête
Corrie
Glacial trough
Hanging valley
Waterfall
Lake
Moraine

Main valley now has deep 'U' shape of **glacial trough**. Tributary channels were less deeply eroded and now form **hanging valleys** (often with a **waterfall** at the point where they meet the main trough). Hollows where ice once collected now appear as steep-sided **corries** (often with a small lake or **tarn**). Ridges cut by glaciers on either side stand out as 'knife-edged' **arêtes**. Floor of trough has **moraines** and possibly a **lake**.

The geological effect of the wind

In most land areas moisture prevents the wind from moving surface material. This moisture has two important effects:

1. It helps to hold loose particles together (have you ever tried blowing moist sand from a dish and then repeated the test with dry sand?);
2. It is essential for plant growth (plants give added protection from wind action by shielding the ground and binding the soil together with their roots).

Wind is therefore only an effective geological agent in dry desert regions where moisture and plants are absent. The speed of the wind governs the size of particles it can move. Although a faint breeze will lift and carry fine dust, small pebbles will only roll slightly in gusts of 150 km/hour. Once the wind speed rises over 20 km/hour sand grains begin to be rolled and bounced along. This movement is called **saltation** and involves grains striking against each other. For example, as one grain bounces it may hit others with enough force to send them bouncing forward as well.

Since saltation involves grains constantly hitting one another, desert sands tend to contain small very-well-rounded grains. These grains are of hard resistant minerals such as quartz because weaker materials break down into dust when bombarded in this way. Grain impacts are much more forceful in wind action than in rivers or seas. This is because there is no water to cushion the collision. Grains striking against bare desert rocks have a sand blasting effect which smooths and shapes the surface. Pebbles polished by this process are known as **ventifacts**.

Wind deposits in deserts

The best-known features are sand dunes which build up according to the prevailing wind direction. Fig. 6.43 illustrates a cresent shaped or **barchan dune** while Fig. 6.44 on page 51 shows the long ridges of sand produced by **seif dunes**. Desert sands may also show **ripple marks** or a type of **cross lamination** commonly called dune bedding (see page 75).

Wind action away from deserts

Strong winds blowing off the sea onto sandy beaches can gradually move the material inland to form coastal dunes. This happens in several south-west facing bays in Britain: for example, Newborough Sands, (Llanddwyn), Anglesey.

Fine dust particles can be carried in the wind for huge distances. Dust from the Sahara, for instance, often travels with the wind and falls in the rains over Europe and the Atlantic Ocean. In fact wind-blown dust is the most common sediment to reach mid-ocean areas.

During the Ice Age, winds were responsible for blowing fine **rock flour** (page 60) far from where melting ice had originally deposited it. This material now forms a fertile cover of **loess** across parts of Europe and the U.S.A.

Wind can also remove soil from farmland, especially when trees and hedges have been removed and ploughing has been carried out during a particularly dry spell. This soil loss is becoming a problem in the large wheat fields of East Anglia.

Finally you should remember that since the wind is responsible for creating sea waves it is indirectly a cause of coastal erosion and deposition.

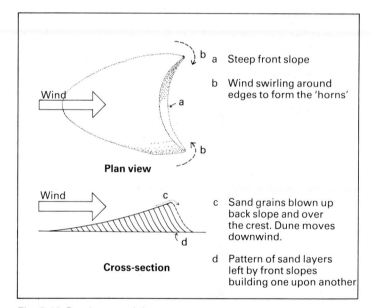

Plan view

a Steep front slope

b Wind swirling around edges to form the 'horns'

Cross-section

c Sand grains blown up back slope and over the crest. Dune moves downwind.

d Pattern of sand layers left by front slopes building one upon another

Fig. 6.43 Barchan sand dune

Fig. 6.42 Experiment to demonstrate transportation and sorting by wind (safety glasses must be worn)

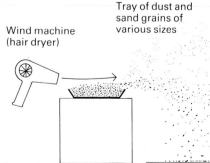

Wind machine (hair dryer)

Tray of dust and sand grains of various sizes

Beam of light to show fine dust staying in suspension (the experiment is best done in a darkened room)

White card to collect deposits: note position of different-sized grains

Questions

For questions 1–4 write the letters (A, B, C...) of the correct answers in your notebook.

1 Which of the following are examples of weathering and which are examples of erosion?
A rocks splitting after a winter of repeated frosts
B waves breaking against a cliff
C pebbles rolling along the bed rock of a river channel
D feldspars decomposing to form clay minerals
E the surface of a rock peeling away after repeated heating and cooling
F wind blown sand grains striking a desert boulder

2 Which **one** of the following would be the most immediate result of a world-wide rise in temperature?
A isostatic uplift of continental shelves
B formation of glacial troughs
C a rise in sea level
D decreasing amounts of volcanic activity
E increasing rates of erosion

3 Which **one** of the following could **not** be formed by river deposition?
A alluvium   B gorge   C sandstone
D delta   E alluvial fan

4 Which **two** of the following factors are **vital** to the formation of a glacier?
A a lack of rivers   B a mountain area
C heavy snowfalls   D steep-sided valleys
E low temperatures all year round

5 Use the following list to choose and write down the term which fits each description (a)-(g) below.
levée / barchan / alluvial fan / till / stalagmite / striation / ventifact
(a) desert pebble polished by being 'sandblasted' in the wind
(b) raised river bank
(c) deposit of $CaCO_3$ formed by water dripping onto a cave floor
(d) deposit of unsorted glacial debris ranging in size from clay particles to boulders
(e) cresent-shaped sand dune
(f) deep scratch left by glacier moving over bare rock
(g) deposit of pebbles and sand formed where a fast flowing stream loses energy at the foot of a slope

6 Name the sedimentary rock types which are most likely to form in the following environments.
(a) beach where pebbles are being constantly rolled together
(b) lower part of continental slope where turbidity currents are acting
(c) coastal lagoon in dry arid climate
(d) clear warm waters on a shallow part of the continental shelf where little sand or mud is being supplied
(e) tranquil offshore part of the continental shelf

(f) abyssal plain onto which deposits of wind-blown dust are occasionally sinking
(g) area of continental shelf, close to shore where current action is relatively powerful.

7 The diagram below shows a cross section through part of a coast.

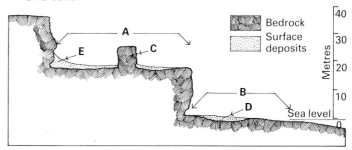

(a) Name the features labelled A and B.
(b) How was feature C formed?
(c) Describe the type of deposit you would expect to find at D.
(d) What has caused deposition at E?
(e) Carefully explain the sequence of events which has produced this coastal shape.

8 Study this map of a coastal area.

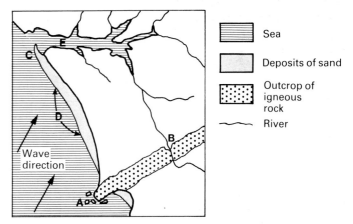

(a) Explain the formation of the headland at A and the gorge at B.
(b) Explain the formation of the sand spit at C.
(c) Groynes have been built on the sandy beach at D; what is the purpose of groynes?
(d) Explain the formation of the large inlet of sea at E; what are inlets like this called?
(e) What type of sediment is likely to be deposited in the shallow waters of this inlet at low tide? Explain your answer.

9 Explain why a clay pit is more likely to become flooded than a sandstone quarry.

10 How true is this statement? 'If the earth had no atmospheric gases or water, there would be no weathering, no erosion and no sedimentary rocks.' Carefully explain your answer.

7 Sedimentary rocks

Chapter 6 explained how weathering and erosion break down rocks to form sediments. At the time of deposition these sediments are described as being **unconsolidated** (a loose mixture of separate fragments). Given time and the right geological conditions, they will become **lithified** (compressed and cemented) into solid sedimentary rocks.

Geological maps, and your own fieldwork, will show that sedimentary rocks are the most common type exposed at the earth's surface. Despite the fact that they cover 75% of the land surface, they actually account for only 7% of the total volume of the crust. In effect they form the 'top covering' of both the continental and oceanic areas.

Sedimentary rocks are classified (divided into groups) according to how they were formed. Two major groups are recognised.

1. Clastic sedimentary rocks. These are formed from material which has been transported as actual pieces of rocks and minerals: for example, pebbles, sand grains, clay minerals, etc. This type of material could be described as the 'visible load' carried by water, ice or wind. The name clastic comes from a Greek word meaning 'broken' and, of course, clastic rocks contain pieces which have been broken from other older rocks.

2. Chemical and organic sedimentary rocks (sometimes called non-clastic). These are mainly formed from material which was transported as a solution dissolved in water (i.e. the 'invisible load'). Either **chemical** reaction or **organic** activity (the actions of living things) can cause the material to come out of solution and form a sediment, for example:

- a deposit of halite (rock salt) can be formed by chemical precipitation as sea water evaporates;
- as sea shells grow they extract dissolved $CaCO_3$ from the water to build their shells, and eventually a deposit of such shells may form a sedimentary rock.

Fig. 7.1 Formation and classification of sediments

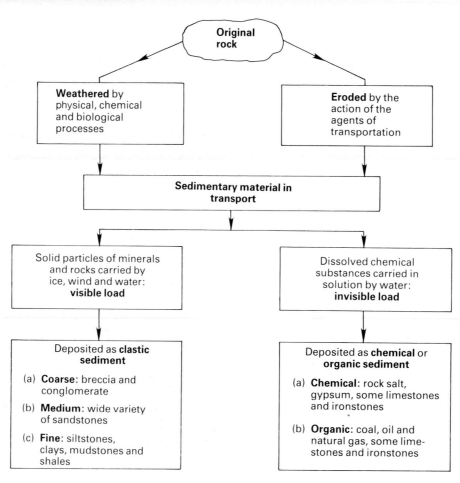

Bedding in sedimentary rocks

Perhaps the most obvious feature of sedimentary rocks, especially in the field, is the **bedding**. As deposits of sediment build up they tend to produce a series of layers. Each layer is called a **bed** and the junctions between them are called **bedding planes**.

Bedding stands out quite clearly when each layer has a slightly different appearance to those above and below it. Even when all the sediment looks very similar it is usually possible to spot the bedding planes produced by pauses during deposition.

Really thin layers of sediment (only a few millimetres thick) are sometimes called **laminae**. Deposits of sediment which do not show bedding are said to be **massive**.

Most sedimentary rocks, particularly those formed on the sea floor, were originally deposited in horizontal beds. In some cases they remain horizontal even after they have been raised and exposed on land (Fig. 7.3). However it is more likely that the uplifting forces also caused some tilting or folding to occur. Beds which are tilted at an angle are said to **dip** (Fig. 7.4). Although most sediments were originally horizontal this is not true in every case. For example, sandstones deposited on a sloping delta front (Fig. 6.18, page 49) may have a slight dip even before any earth force has affected them.

As a series of beds is deposited one on top of another they record a time sequence of geological events. The oldest bed is normally at the bottom of the series and the youngest at the top. Occasionally, however, exceptionally powerful folding can completely overturn a series. We shall explain how to recognise such situations later (page 97).

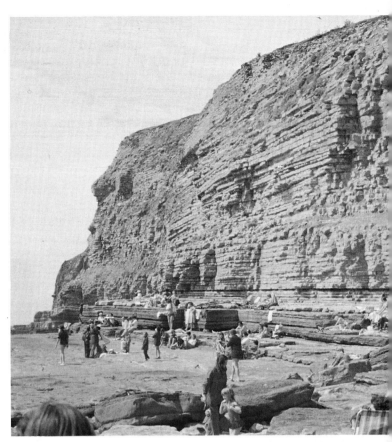

Fig. 7.3 Horizontal beds of sedimentary rock, near Porthcawl, South Wales

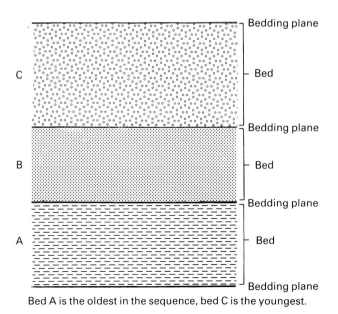

C — Bedding plane
C — Bed
— Bedding plane
B — Bed
— Bedding plane
A — Bed
— Bedding plane

Bed A is the oldest in the sequence, bed C is the youngest.

Fig. 7.2 A bedded sequence of sedimentary rocks

Fig. 7.4 Dipping beds of sedimentary rock

65

Clastic sedimentary rocks

These sediments are formed from broken fragments of older rocks. The size of the fragments determines how they can be transported and where they are deposited. It is therefore convenient to divide clastic sediments into three groups according to their grain size.

- **Coarse grained.** These contain fragments over 2 mm in diameter and include the rock types **conglomerate** and **breccia**.
- **Medium grained.** These are the **sandstones** and contain fragments of between 2 mm and 0.05 mm diameter.
- **Fine grained.** These contain tiny fragments of less than 0.05 mm diameter and include the rock types **siltstone, clay, mudstone** and **shale**.

Coarse-grained and medium-grained clastic sediments usually contain some additional material which fills in the spaces between their fragments. This infilling has the effect of holding the sediment together as a solid rock. There are two types of infill material.

1. Matrix. This term is used for fine silt or clay which has been deposited together with the larger fragments. As the sediment is compacted the fine matrix particles tend to 'fuse' and bind the deposit together.

2. Cement. This term is used when minerals have been precipitated into the spaces between fragments. After deposition it is common for water to seep through a sediment. Given the right conditions (which often occur in the warmth and pressure after the sediment is buried) substances such as $CaCO_3$, $Fe(OH)_3$ and SiO_2 will precipitate and 'glue' the deposit together. Note that the term 'cement' is used in geology to describe any mineral which binds sediment into solid rock; it does not refer to the making of concrete or mortar.

Coarse-grained clastic sedimentary rocks

Although these rocks are defined as having fragments greater than 2 mm in diameter, many contain pebbles or pieces of material well above this minimum size. The two main rock types within this group are distinguished by the shape of their fragments.

Conglomerates (see Fig. 7.5, page 51)
Conglomerates are rocks with large rounded fragments (pebbles). In some samples you will see that the pebbles are all of a similar size, all very well rounded and all made of resistant material such as quartz or metaquartzite. These probably formed under beach conditions. The constant wave action will have broken up and removed weaker rock fragments

leaving the resistant ones to tumble against each other and become rounded. Other examples of conglomerate may have a mixture of pebble sizes and types which are all less well rounded. Such rocks were possibly formed where a river deposited a thick alluvial fan (page 48). An extra clue to this origin would be a high proportion of sand grains deposited between the pebbles. Perhaps 'pebbly sandstone' is a better term for rocks of this type.

When examining conglomerates it is worth looking to see if the pebbles are aligned. This can help you to work out which way the original current was flowing.

Breccias (see Fig. 7.5, page 51)
Breccias contain large angular fragments. Such material could not have been transported far or rounding would have occurred. The presence of a mixture of rock types in a range of sizes also suggests rapid deposition (no time for fragments to be sorted). Some breccias are formed, for example, when sudden rainstorms cause **flash floods** in deserts. Large rock fragments are rapidly transported by the water and deposited into gullies. Other breccias are formed from **screes** (see page 43). *Note:* fault breccias are formed by different processes and are classified as metamorphic (see page 82).

You will probably examine some specimens and discover both rounded and angular fragments in the same rock. This is to be expected where a range of rock types were eroded but during transport only the weaker types became rounded. Such specimens are called **breccio-conglomerate**.

Medium-grained clastic sedimentary rocks

This group is defined as having fragments between 2 mm and 0.05 mm in diameter. It contains the type of rocks commonly known as **sandstone**.

Sand grains can be transported by wind, ice or water and deposited in deserts, rivers, lakes or the sea. With such a variety of origins it is no wonder that many different types of sandstone occur. When studying these rocks, consider the following points; they will help in your descriptions and identifications.

1. Grain size. You can recognise sand-sized particles quite easily. Although they are less than 2 mm across they are still visible to the naked eye. Sandstones also feel rough to the touch.

2. Grain shape. Use a hand lens to see if the grains are rounded or angular. Angular grains are described as 'grit', and rocks containing them have a sharp gritty feel.

3. Grain sorting. Sandstones are said to be well sorted if all the grains are approximately the same size, and poorly sorted if a wide range of sizes is present. (See practical work on sorting, opposite.)

4. Mineralogy of grains. Grains of quartz are the commonest material in most sandstones. They can be recognised by their hardness and glassy appearance. Other minerals include feldspar and silvery flakes of muscovite mica. Occasionally, rock fragments such as small pieces of slate or basalt may be present.

5. Type of cement. Common cements are silica (a form of quartz, recognised by its hardness), calcite (recognised by the acid test) and iron compounds (identified by their rusty red colour). The strength of the cement governs how easily grains may be broken away from the rock. Weak poorly cemented sandstones are said to be **friable**.

Fig. 7.10 on page 51 shows a microscope photograph of sandstone. How would you describe the texture of this rock?

Before studying sandstone rocks it is useful to investigate some of the features of loose sand deposits. You can collect these from river banks, lake sides and beaches.

Experiment 1: Investigating grain size and sorting.

For this experiment you will need a set of sieves like those shown in Fig. 7.6. Each sieve has a different mesh size and they are stacked up with the coarsest mesh at the top and the finest at the bottom. At the base is a pan to collect the very fine material which has passed through the bottom sieve.

Before sieving a sand sample make sure that:

- the sand is absolutely dry,
- the grains are not sticking together (crumble any lumps between your fingers);
- the sample has been accurately weighed (about 100 to 150 g is a suitable amount).

Place the sample in the top sieve, put on the lid and gently shake the stack for 15 minutes. Make sure no

Fig. 7.6 Set of sand sieves

sand is spilt. When shaking is finished, you will have separated the sample into a number of different grain sizes. Weigh the sand collected in each sieve and calculate what percentage it is of the total weight of the whole sample. For example,

Percentage of sample in sieve 1

$$= \frac{\text{weight of sand in sieve 1}}{\text{total weight of sample}} \times 100\%$$

Plot your results on a bar graph. You should repeat the process for different sand samples but make sure you clean and brush each sieve between experiments.

Fig. 7.7 shows graphs from two actual samples. *Which would you describe as well sorted, and which as poorly sorted?*

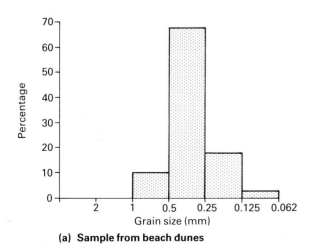

(a) Sample from beach dunes

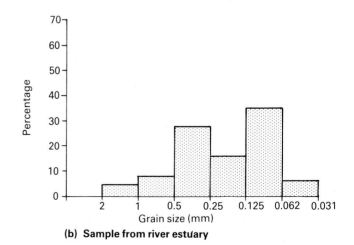

(b) Sample from river estuary

Fig. 7.7 Bar graphs showing grain size of sand deposits

Experiment 2: Investigating grain type. This experiment is best done using sand grains of between 1 mm and 2 mm in diameter. Select, at random, 100 grains from your sample. Then, using a magnifying glass, try to identify them. Look particularly for quartz, feldspar, mica and calcite grains (use the mineral table on page 25 as a guide). You will probably also find some tiny rock fragments. *When you have divided your 100 grains into suitable categories, count how many are in each, and draw a bar or pie graph of the result.*

You can repeat this task using a sample of beach pebbles or river gravel. These fragments of rock may prove more difficult to identify and classify than single grains of minerals. You could follow up this study by looking at a geology map of the area where your sample comes from. You may be able to work out where the sediment has been eroded from.

Main types of sandstones (see Fig. 7. 9, page 51)

Geologists divide sandstones into five main types according to their mineralogy and texture.

Orthoquartzite (quartz sandstone)

As the name suggests these sandstones are made almost entirely of quartz grains. Fig. 7.8a gives a microscope view which clearly shows that these rocks are well sorted and contain well rounded grains.

A typical environment for an orthoquartzite would be a shallow sea where current action constantly reworks the grains by rolling them back and forth. Deposition tends to be slow so there is plenty of time for weaker minerals such as feldspar to be broken down completely rather than buried in the sediment. Only the strong quartz grains remain and they become rounded and sorted as the movement continues. The current may also produce **ripple marks** and **cross lamination** (page 75).

Orthoquartzites are especially likely to form where a supply of rounded quartz grains is arriving from the erosion of older sandstone rocks. Most orthoquartzites are cemented by quartz which crystallised to fill the spaces between the grains.

Arkose

These sandstones contain a high proportion of **feldspar** grains (typical composition 65% quartz, 35% feldspar) and have a pinkish colour. They tend to form where sediment is supplied from feldspar-rich rocks (for example, granite and gneiss) which have been mainly affected by physical weathering. (Chemical weathering would cause the feldspar to decompose and form clay minerals.) Rapid deposition is also required to prevent the feldspar being broken down during transportation. Because of their poorly sorted angular texture (Fig. 7.8b), arkoses have a gritty feel.

Greywacke

This odd name actually means 'grey grit'. Greywackes are impure sediments with a high proportion of fine **clay matrix**. The sand-sized grains are angular, poorly sorted and of mixed type (Fig. 7.8c). Greywackes are formed when a range of clastic sediment is dumped so quickly that there is no chance for proper sorting to take place. **Graded bedding** (page 76) is a common feature of greywackes, which suggests **turbidity currents** are often responsible for their deposition.

'Desert' sandstone

Sandstones formed in desert environments of the past are quite common in Britain. A microscope view of one of these rocks is shown in Fig. 7.10 on page 51. Since the rock is made almost entirely of quartz grains, desert sandstone is really a special form of orthoquartzite. However, a distinctive feature is the red/brown cement of iron minerals such as **haematite**. This cement is rather weak, so desert sandstones tend to be **friable**. Outcrops of desert sandstone often show large-scale **cross-lamination** because the sand built up into dunes (see Fig. 7.26 and page 62).

Fig. 7.8 Microscope views of sandstones

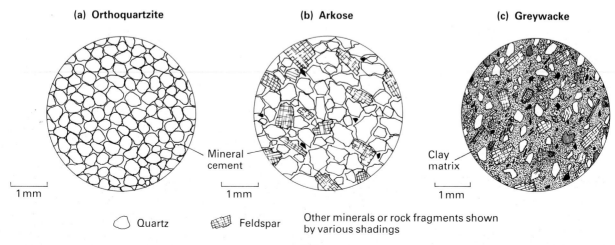

(a) Orthoquartzite **(b) Arkose** **(c) Greywacke**

Mineral cement

Clay matrix

1 mm 1 mm 1 mm

◯ Quartz   ▦ Feldspar   Other minerals or rock fragments shown by various shadings

Micaceous sandstone

These sandstones contain flakes of **muscovite mica** deposited parallel to the bedding. The presence of this mineral allows the rock to be split into sheets (usually a few centimetres thick). Such rocks were once used as paving slabs and are often called **flagstones**.

Fine-grained clastic sedimentary rocks

(See Fig. 7.9, page 51)

These sediments are made of tiny particles less than 0.05 mm in diameter (that is, individual grains are too small to be seen with the naked eye).

The smallest and finest quartz grains are included in this category. They form rocks known as **siltstones**. A siltstone can be distinguished from a sandstone by its smoother feel (it feels similar to the finest grades of abrasive paper, like those used for rubbing down car paintwork). Many siltstones show exceptionally thin bedding which is called **lamination**.

Materials even finer than silt are deposited in quiet low-energy conditions. These are the clay minerals, formed during chemical weathering (page 43), and carried in suspension by moving water. Clay minerals belong to the mica family and each tiny crystal (about 0.005 mm across) has a flat plate-like shape. Several varieties of rock can be produced by the deposition of clay minerals.

Clay

This material is easily recognised since, when wet, it can be moulded like Plasticine. It occurs where deposits of clay minerals have not been compressed or hardened into more solid types of rock. Most clays are fairly recent deposits (for example, those formed in glacial lakes during the Ice Age). But, surprisingly, others have survived for millions of years (for example, the Oxford Clay, excavated for brick making at many sites between Oxford and Peterborough, was laid down 160 million years ago). Other economic uses of clay are mentioned on page 185.

Fig. 7.12 Diagram showing the different arrangement of clay minerals in mudstone and shale (greatly magnified)

(a) Shale (parallel)

(b) Mudstone (random)

Shale

When clay minerals are deposited slowly (for example, in deep quiet waters), they settle from suspension and come to rest with their flat 'plates' lying horizontally. During burial, pressure from above squeezes the sediment and aligns the minerals even more (see Fig. 7.12a). The resulting rock is called shale. Since it is made of layers of clay minerals, shale is easily broken into thin flat slices and is said to be **fissile**.

Many black shales contain the remains of plankton (minute plant and animal life found in sea water). These rocks probably formed in deep or exceptionally still water where oxygen was lacking. In these conditions no creatures can survive at the sea bed, so any plankton that sinks is preserved because there is nothing there to feed on it and destroy it.

It is important not to confuse shale with **slate**. Slate is a metamorphic rock (page 84) and splits into much stronger and harder sheets.

Mudstone

This rock is produced when clay minerals are deposited more rapidly and collect in a 'jumbled' fashion (see Fig. 7.12b). Provided there is no great pressure during burial the clay minerals do not become aligned so the rock is not fissile and breaks unevenly. If you have difficulty deciding whether a specimen is siltstone or mudstone, rub it gently across your teeth! A grating feeling indicates siltstone, while a smooth feeling indicates mudstone.

Fig. 7.11 Brickworks on the Oxford Clay at Ridgmont, Bedfordshire

Chemical and organic sedimentary rocks

This group includes sediments produced by chemical precipitation and/or the action of living organisms (see page 64 for a full definition).

Limestones (carbonate sediments)

(see Fig. 7.13, page 51)

Limestones are sedimentary rocks which contain **calcite** ($CaCO_3$) as their main mineral. They are easily identified because bubbles of carbon dioxide are given off when a drop of dilute hydrochloric acid is placed on them.

Calcium carbonate is transported dissolved in water. Limestones may be formed by direct precipitation and/or the accumulation of deposits of shell material.

Shelly limestone

Many invertebrate animals (animals with no backbones) take $CaCO_3$ from sea water and use it to build shells or skeletons. In areas of clear water (where hardly any sand or mud is being carried in) the remains of these shells and skeletons may be practically the only sediment being deposited. Shelly limestones are not normally made of whole shells because current action reduces much of the material to smaller fragments. The most commonly recognised fragments come from **brachiopods, bivalves, gastropods** and **cephalopods**. Some limestones may contain fragments of **crinoid stems** (see Fig. 12.32, page 128). For further information about these fossil types see Chapter 12.

Shelly limestones are sometimes called **bioclastic limestones** (bio means 'living'; clastic means 'broken').

Fig. 7.14 Coccoliths (greatly magnified by an electron microscope)

Chalk

Chalk is a white very pure (98% $CaCO_3$) form of limestone. Although a hand specimen shows no obvious fossil remains, a powerful electron microscope reveals that it is made almost entirely of minute calcite discs like those shown in Fig. 7.14. Each disc is the skeleton of a tiny plant called a **coccolith** which lived floating in the plankton of sea water. Since each coccolith is only a few thousandths of a millimetre in diameter it seems almost incredible that they could accumulate in sufficient numbers to form the thick beds of chalk found, for instance, across a wide area of Western Europe. We shall return to the unusual conditions needed for chalk formation when we discuss the geological history of Britain (see page 167).

Reef limestone

Given suitable conditions of clear warm sea water, colonies of **coral** and other living organisms may grow to form solid **reefs**. Reefs are strong rock-like structures, firmly fixed to the sea bed (ships can be wrecked against them in such places as the Great Barrier Reef of Eastern Australia). Reefs from ancient seas are preserved as hard reef limestone, often standing out as small hills from surrounding beds of shelly limestone.

The exact conditions of reef formation are dealt with in the section on **corals** (see page 130).

Lime 'mudstone' (micrite)

In certain situations, $CaCO_3$ may precipitate directly from sea water. This is particularly likely in areas of warm rather tranquil waters where some evaporation is taking place. A typical environment would be a flat shallow 'bank' where current action is weak.

The precipitation forms tiny crystals which sink to produce a fine white 'mud' called **micrite**. Sometimes ordinary mud (made of clay minerals) is deposited with this material to give the final rock a dark grey colour. You should always test mudstone specimens with acid to see if they contain any precipitated lime.

Micrite may also be found as a matrix in some shelly limestone.

Oolitic limestone

Because lime 'mud' is a very sticky substance it can form a coating on other fragments of sediment. Where mild current action rolls silt grains or tiny pieces of shell they may develop into small lime-coated spheres called **ooliths**. As Fig. 7.15 on page 52 shows, each oolith is usually less than a millimetre in diameter and has developed by a snowball effect as layer upon layer of lime coated its surface. Modern sediments of this type are presently forming on the Bahama Bank off the Florida coast.

Fig. 7.16 Coal forest: a reconstruction of how parts of Britain may have looked during the Carboniferous period

Coal

Coal is a sedimentary rock formed from the remains of trees and plants (vegetation). You may think it should not be included with the sediments since it appears to contain no 'second hand' material released from other rocks by weathering or erosion. However, remember that plants grow by extracting chemical elements from the soil, water and atmosphere, and these elements have originally come from the breakdown of rocks.

Plants also grow by absorbing energy from the sun's rays. They use it in the process of photosynthesis which combines H_2O and CO_2 to make cellulose (plant tissue). Coal is described as a fossil fuel because it contains energy trapped by living organisms of the past.

Most plant material simply rots away after death. Special conditions are needed for it to be preserved and altered into coal. The first requirement is for there to be plenty of vegetation so that a thick deposit can build up. This is most likely to occur in a hot wet tropical climate. A second need is for the plant material to accumulate in boggy conditions where little oxygen is available in the almost stagnant water. Without oxygen, bacteria cannot successfully break down and rot the plant tissue. Thirdly, the deposit of unrotted vegetation must eventually be covered by some other sediment such as sand or mud. This makes sure that burial and long-term preservation can occur.

Most of the coal deposits of Western Europe and North America were originally formed about 300 to 280 million years ago during a period of time called the **Carboniferous**. The climate was then ideal for large 'coal forests' to flourish (Fig. 7.16). In addition, there were huge areas of low-lying swampy ground at this time because a series of deltas was then infilling a shallow sea. We shall return to the story of how Carboniferous forests grew, sank and were covered by delta sediments later in the book, when we study the geological history of Britain.

After burial, the vegetation is squeezed and compacted by the weight of sediments above. Chemical changes take place which slowly turn a bed of soft brown woody material into a seam of hard black coal. The main change involves the breakdown of cellulose by reactions of the following type:

$$C_6H_{10}O_5 \rightarrow CO_2 + 3H_2O + CH_4 + 4C$$

(Cellulose) (Carbon dioxide) (Water) (Methane) (Carbon)

Not all the cellulose reacts in this way. Instead, there is a gradual change and, as the water and gases escape, the remaining material becomes increasingly richer in carbon. A range of carbon fuels called the **coal series** is produced according to how much change has taken place (see Fig. 7.17). In general, the longer burial lasts or the deeper it goes, the more carbon is concentrated and the better the fuel becomes.

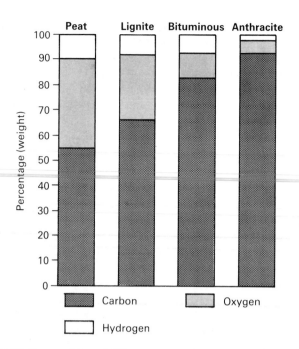

Fig. 7.17 Composition of different types of coal

Lignite

This represents the next stage where more of the cellulose has been altered to carbon. It is dark brown, still shows some plant remains but burns better than peat. The thick beds of lignite mined in East Germany were deposited some 200 million years later than the Carboniferous coals found elsewhere in Europe. This may explain why they have not been changed into such good fuels.

Bituminous coal

This is the type of Carboniferous age coal commonly mined in Britain. It it fairly hard, has a shiny black colour and burns well. Estimates suggest that 10 to 15 metres of vegetation have been compacted to produce each metre of bituminous coal.

Anthracite

This is the best coal of all. It has a high carbon content, burns with a hot bright flame and can be recognised by its **conchoidal (shell-like) fracture**. Anthracite is mined in parts of the South Wales coalfield where high temperatures have been responsible for its formation.

The extent of Britain's coal deposits and the methods of mining them are dealt with in the chapter on economic geology (page 181). That chapter also explains the origin of other fossil fuels such as petroleum and natural gas.

Peat

This burns with a lot of smoke and gives a rather poor flame. It is soft, brown and loose with plenty of visible plant remains. Most peat has only been deposited fairly recently and so has had no real chance to be compacted and altered to proper coal. It is burned where other better fuels are not available, for example in Southern Ireland and the north of Scotland.

Fig. 7.18 Peat cutting, Achill Island, Ireland

Evaporite sediments

As the name suggests, these sediments are formed when water evaporates and precipitates its dissolved chemicals. Although some rare evaporite deposits are formed when fresh water lakes dry up, much larger amounts come from sea water. The chief marine evaporites are **halite** ($NaCl$) and **gypsum** ($CaSO_4$ $2H_2O$).

Sea water contains very small quantities of most elements in solution but, as Fig. 7.19 shows, only a few substances are present in reasonably large amounts. As evaporation begins, the least soluble chemical is the first to be precipitated. In sea water this is calcium carbonate which forms micrite (limestone) in the way described on page 70. The next material to precipitate is gypsum but before this can happen over 80% of the water must be evaporated. Halite follows when 90% of the water has gone, and finally the most soluble potassium (K) and magnesium (Mg) compounds come out of solution when only a small percentage of the original sea water is left.

Collect several litres of sea water and evaporate it to see what materials are precipitated.

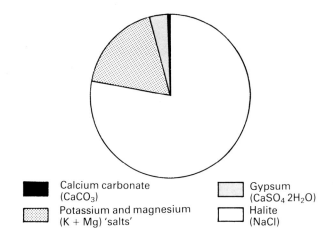

Calcium carbonate
(CaCO₃)

Potassium and magnesium
(K + Mg) 'salts'

Gypsum
(CaSO₄ 2H₂O)

Halite
(NaCl)

Fig. 7.19 Relative amounts of dissolved 'salts' present in sea water. (In total these 'salts' make up 3.5% of the volume of sea water)

Although $CaCO_3$ may be precipitated in the waters of an open sea, the rate of evaporation there is never high enough to allow gypsum or halite to form. These minerals only come out of solution in shallow lagoons or inland seas, which evaporate in a hot dry climate. Even then, the amounts produced are quite small; over 3 metres of water must be evaporated to produce just 5 centimetres of evaporite minerals.

Sometimes, however, thick deposits of evaporites can form: for example, there is a 1000 m sequence at Stassfurt in Germany. The 'bar theory' (Fig. 7.20) has been suggested as an explanation of this. If a shallow coastal lagoon has a slightly raised edge (bar) separating it from the open sea, the lagoon is constantly 'topped up' by water flowing over the bar at high tide. Each new 'top up' is trapped behind the bar and, as evaporation takes place, minerals are precipitated

onto the floor of the lagoon. Given enough time, a thick sequence can eventually build up, particularly if the floor of the lagoon is subsiding under the weight of the evaporites. This explanation is really more than a theory because it is known to be happening at the present time. For example, evaporites are now forming in the Gulf of Kara Bugaz in southern U.S.S.R. This shallow body of water has a narrow link (over a 'bar') with the Caspian Sea.

In some cases evaporites are precipitated on broad coastal salt flats called **sabkhas** (pronounced 'sabkas'). Instead of being permanently covered by a lagoon, such places are only flooded occasionally by the sea which then evaporates as the region dries out once more. Present-day sabkhas are to be seen along the coastline of the Persian Gulf.

Gypsum

This white, very soft mineral (hardness of 2 on Mohs' scale) is mined as a source of plaster. Plaster of Paris gets its name because of the gypsum found close to that city. In Britain there are deposits in Nottinghamshire and near Appleby in Cumbria.

Halite

This mineral is commonly called rock salt and is easily recognised by its taste. British deposits are found in Cheshire and Teesside where (like the gypsum beds mentioned above) they occur as part of a sequence of desert sandstones. Salt is used in the chemical industry, in food preparation and as a de-icer for winter roads.

(a)

(b)

Fig. 7.21 Specimens of (a) gypsum and (b) halite

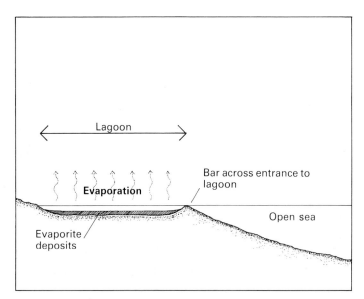

Fig. 7.20 The bar theory of evaporite deposits

73

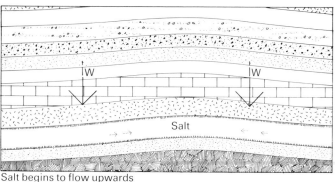
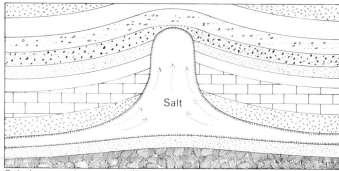

Salt begins to flow upwards
W Weight of overlying sediments

Salt dome pushes up and through overlying sediments (rather like an intrusion)

Fig. 7.22 Formation of a salt dome (the diagram represents a salt bed buried to a depth of several thousand metres)

An unusual feature of halite deposits is their ability to become semi-plastic and flow under pressure. This tends to happen when salt beds are buried to depths of between 3000 and 4000 metres. Because halite is much less dense than other rock types, the flow moves upwards to produce a **salt dome** (see Fig. 7.22). Salt domes are one type of structure that provides a trap for oil and natural gas (see page 182).

Magnesium and potassium evaporites

These chemicals are so soluble that only exceptional amounts of evaporation can cause them to be precipitated. A deposit of **potash** (potassium/magnesium chloride) is mined at Boulby in North Yorkshire for use by the chemical and fertiliser industries.

Sedimentary ironstones

Iron is a common element in the earth's crust and a variety of iron compounds are released into rivers and oceans by chemical weathering. If conditions such as water temperature and acidity are suitable, the iron may come out of solution to form a sediment on a lake or sea bed. The action of bacteria can sometimes help this process.

Beds of sedimentary ironstone are found in Northamptonshire, Lincolnshire and Yorkshire. Despite being of rather poor quality the ores have been used because they are close to the surface and easy to quarry. The main iron minerals present are **limonite** ($FeO(OH)$), **siderite** ($FeCO_3$) and **chamosite** (complex iron silicate).

It is likely that some ancient ironstones were produced because the chemistry of the oceans and atmosphere were different during those past times. For example, the 2000 million year haematite beds of Canada probably formed in an environment where there was little atmospheric oxygen (see Fig. 2.6 on the evolution of the atmosphere).

Fig. 7.23 Open cast ironstone mining, Lincolnshire

Sedimentary structures

Although the most obvious feature of sedimentary rocks is usually the **bedding** (page 65), on closer examination you will often discover other structures which give additional clues about the style of deposition.

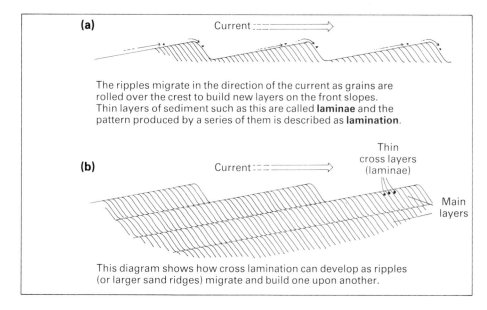

(a)

Current

The ripples migrate in the direction of the current as grains are rolled over the crest to build new layers on the front slopes. Thin layers of sediment such as this are called **laminae** and the pattern produced by a series of them is described as **lamination**.

(b)

Current

Thin cross layers (laminae)

Main layers

This diagram shows how cross lamination can develop as ripples (or larger sand ridges) migrate and build one upon another.

Fig. 7.24 Ripples and cross-lamination

Ripples

Ripples are formed by currents of water or air flowing over the surface of a sediment. No doubt you have seen them on sandy beaches. Fig. 7.24a illustrates the typical pattern. Notice how each ripple moves forward with the current. You may be able to watch this happening in shallow pools or try an experiment with a low-powered electric fan blowing across a tray of dry sand (wear safety glasses).

Ripples do not only occur on beaches; they can form wherever currents flow across sea, lake or river beds. Their size is very variable and in some cases they take the form of parallel sand ridges up to several metres high. Desert sand dunes of the type shown on page 62 are really a special case of giant ripples being created and moved by the wind.

Finding ripple marks preserved on a bedding plane allows you to work out which way the current was moving. Look for them particularly in sandstone exposures.

Cross lamination

This type of structure develops from the formation and movement of ripples (see Fig. 7.24b). Its main feature is a series of angled layers of sediment within the main sedimentary beds. The angle, pattern and width of these layers depend on the style of their deposition. Fig. 7.25 shows small-scale cross lamination produced by current action in shallow water. Fig. 7.26 shows an exposure of desert sandstone with the giant ripple effect of sand dunes clearly visible. Other variations of cross lamination exist. Many show evidence that the tops of some ripples were eroded before the next set was deposited above. The only real way to learn about these is by first-hand examination of specimens.

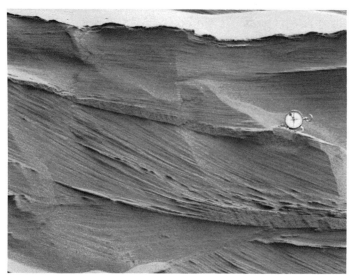

Fig. 7.25 Small-scale cross lamination

Fig. 7.26 Large-scale cross lamination (dune bedding) seen in desert sandstones near Dawlish, Devon

Graded bedding

This term is used where grain size decreases from the lower part of a bed to the upper part of a bed (see Fig. 7.27). You can easily produce this structure by mixing sand, silt and clay with plenty of water then allowing the mixture to settle in a plastic beaker. *Why do the particles arrange themselves in this order?* You should also look out for graded bedding in fieldwork exposures. You can use it to decide whether a sediment is the right way up (see page 97).

If you think about graded bedding it becomes obvious that rather unusual conditions of transportation and deposition must be responsible for it. Instead of the normal situation (where different grain sizes become sorted from each other and deposited in different places) a mixture of sizes must all be transported to the same place and allowed to settle together.

Many graded beds are deposited by **turbidity currents**. Page 58 explained how these currents develop on continental slopes. A cloud of sediment becomes dispersed in water and moves rapidly downslope, picking up more material (probably of a finer grain size) as it travels. At the bottom of the slope energy is lost and the cloud of mixed sedimentary material settles out to form a graded bed.

Graded beds are not only found along the lower parts of continental slopes. They can also occur where turbidity currents have swept down other unstable slopes such as delta fronts (see Fig. 6.18 on page 49). A deposit produced by turbidity current action can be called a **turbidite**. Most of these rocks

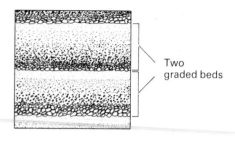

Fig. 7.27 Graded bedding

are actually the 'mixed' type of sediment known as **greywacke sandstone** (page 68).

Fig. 7.28 shows an experiment for making your own turbidity currents and graded beds. The experiment is best done in a fairly long aquarium tank. The water in the tank must be absolutely still before the sediment mix is poured in. Repeat the experiment several times to note the effect of turbidity currents on sediment that has already been deposited (let the water settle between each 'current'). *Note*: if you only wish to view the current, a strong solution of salt (dyed with potassium permanganate) will work instead of the sediment mixture.

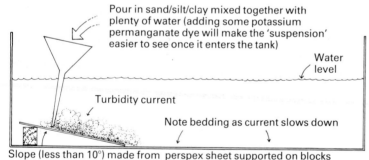

Fig. 7.28 Experiment to demonstrate turbidity current and graded bedding (see text for full details)

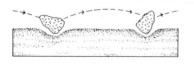

(a) Bed of sediment is 'marked' by pebble bouncing across it

(b) Marks are infilled and preserved by next layer of sediment

Fig. 7.29 Sole marks (this shows one possible way in which the original marks are formed)

Sole marks

Fig. 7.29 shows how small dents or marks in the top of a sediment layer can be infilled and preserved as the next bed is deposited above. Marks can be made by things like pebbles, shells or even clumps of seaweed bouncing or rolling along with the current. Alternatively, sole marks may be formed by turbulent currents scouring soft sediment or (as in Fig. 7.30) where part of the upper sediment has 'collapsed' into the lower one. Other sole marks include infilled **trace fossils** such as worm tubes or dinosaur footprints.

The name 'sole mark' comes from the fact that they are preserved on the bottom of the sediment which infilled them (sole means 'the underneath of', as in the sole of a shoe).

Fig. 7.30 Sole mark preserved on the junction between beds of siltstone and shale

Fig. 7.31 Dessication (sun) cracks in a dried-up lake bed. Note also the small circular marks produced by a short shower of rain

Dessication cracks

These structures are fairly common in sediments from lakes, lagoons or sabkhas which dried out from time to time (dessication means 'drying out'). As Fig. 7.31 shows, such cracks have a polygonal (many-sided) pattern, which is caused by contraction.

You can experiment in their formation by making a thick wet clay/silt mixture and leaving samples of it to dry out in a variety of conditions: for example, on a window sill, radiator top, etc. Try the experiment with wet sand also to see what effect dessication has on slightly coarser-grained sediment.

Lithifaction

Lithifaction literally means 'turning to stone'. Loose sediments are lithified into solid sedimentary rocks by a number of processes.

Compaction. As sediment accumulates, the upper layers press down on the lower ones. This compaction squeezes out the water and packs the grains closer together. Under such conditions some minerals partly recrystallise and effectively lock the sediment into a solid mass (Fig. 7.32 shows this effect on quartz sand grains). The fine **matrix** (see page 66) of some sediments can also harden and bind fragments together during compaction.

Cementation. Water seeping through sediment often precipitates mineral cement between the grains (see page 66). Although the mineral may have been present in solution before the water entered the sediment, this is not always so. In some cases the water has actually dissolved the material from one part of the sediment and then precipitated it elsewhere in the same bed. Cementing minerals may be concentrated at certain points where chemical conditions favour their precipitation: for example, around decaying organic matter or where irregularities such as sole marks occur in the sediment. The concentration of minerals at particular points produces **nodules** such as those illustrated in Fig. 7.33 on page 52. This picture shows nodules of iron minerals but other common examples include calcite nodules in mudstone (test with dilute acid), **flint** nodules in chalk and **chert** nodules in limestone (flint and chert are both forms of silica).

Note that compaction and cementation do not act independently; it is their combined effect which lithifies sediment. Note also that the word **diagenesis** can be used as an alternative term for lithifaction.

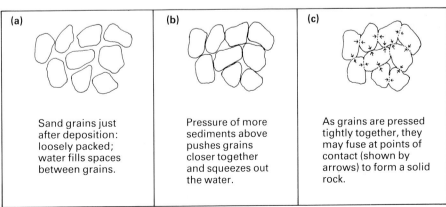

(a)	(b)	(c)
Sand grains just after deposition: loosely packed; water fills spaces between grains.	Pressure of more sediments above pushes grains closer together and squeezes out the water.	As grains are pressed tightly together, they may fuse at points of contact (shown by arrows) to form a solid rock.

Fig. 7.32 Compaction of sandstone. *Note:* compaction decreases the porosity (the degree to which water can be held in a sediment)

Description of sedimentary rocks

When describing sedimentary rocks always try to give as much detail as possible. In addition to studying mineralogy and texture, note features such as bedding, sedimentary structures and the presence of fossils. All this information is useful when working out what type of environment the rock could have been deposited in.

Typical description of a sedimentary specimen

The following description refers to the specimen shown in Fig. 7.34 on page 52.

Texture
- The individual grains do not interlock. Therefore this rock has a fragmental texture.
- Individual grains/fragments vary in size from less than 1 mm to (in a few cases) over 4 mm in diameter. Therefore this rock is poorly sorted.
- Despite the presence of some larger grains, most of this rock could be described as medium grained.
- The specimen has a gritty feel and (under a hand lens) most grains appear to be rather angular.
- Individual grains are difficult to break away from the specimen. Therefore the rock seems well compacted and cemented.

Mineralogy
- Most grains (including all the larger ones) are glassy looking and harder than steel. Therefore these are grains of quartz.
- About 20% of the rock is made up of yellowish-white grains which crumble easily. These are most probably grains of feldspar which have begun to decompose as the rock is weathered.
- The rock does not react with dilute HCl. Therefore no $CaCO_3$ is present in either the grains or cement. However, the acid quickly soaked into the specimen showing that this rock is porous.
- The brown coloration of this rock suggests that iron minerals may form part of the cement.

Other observations
- No bedding planes could be seen in this small specimen.
- No fossils were present.

Conclusion
The fragmental texture shows that this is a clastic **sedimentary rock**. Its texture and mineralogy lead to the conclusion that it is a **gritty sandstone** (of the **arkose** type) which has been rapidly deposited.

Questions

For questions 1–6 write the letters (A, B, C, ...) of the correct answers in your notebook.

1 Loose deposits of sediment are described as being
  A lithified   B eroded   C porphyritic
  D unconsolidated   E cemented

2 Which **two** of these rocks are formed by organic action?
  A chalk   B shale   C coal
  D breccia   E siltstone

3 Which **one** of the following features would you **not** expect to see in an orthoquartzite?
  A well-rounded grains   B well-sorted grains
  C a quartz cement   D numerous feldspar grains
  E bedding planes

4 The typical colour of desert sandstones is
  A greenish blue   B reddish brown   C pale grey
  D white   E very faint green

5 Which **two** of the following statements about clay minerals are false?
  A They are formed by the physical weathering of quartz.
  B They can be carried in suspension by moving water.
  C They are silicate minerals.
  D Deposits of them can form mudstone or shale.
  E They have acicular ('needle-shaped') crystals.

6 Coal was formed in
  A the desert   B deep mines   C the sea
  D glacial environments   E tropical swamps

7 Use the following list to choose and write down the rock type which fits each description (a) –(i) below.
breccia / lignite / siltstone / conglomerate / shale greywacke / chalk / oolitic limestone / arkose
  (a) sandstone with a relatively high proportion of feldspar grains
  (b) rock containing many rounded pebbles
  (c) rock composed of microscopic coccoliths
  (d) brown rather poor quality coal
  (e) type of sandstone with a pooly sorted mixture of mineral fragments in a clay matrix
  (f) fissile fine grained rock made up of clay minerals
  (g) rock containing concentric spheres of $CaCO_3$
  (h) rock formed by rapid deposition of large angular clastic fragments
  (i) rock made up of rounded tiny grains of quartz

8 The diagram below shows a sequence of sedimentary rocks.

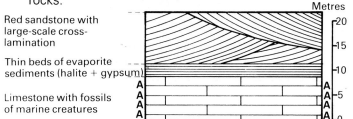

Red sandstone with large-scale cross-lamination

Thin beds of evaporite sediments (halite + gypsum)

Limestone with fossils of marine creatures

78

(a) What thickness of sandstone is shown?

(b) What name is given to the planes labelled A – A in the limestone?

(c) What tests would you use to distinguish between specimens of halite and gypsum?

(d) The limestone contains micrite; what is this material and how is it formed?

(e) The sandstone is described as being friable; what does this term mean and what does it tell you about the type of cement in this rock?

(f) Explain the origin of the type of bedding shown by the sandstone.

(g) Carefully explain how this sequence of sediments could have been formed (your answer should describe how the environments of deposition were changing as time passed and each rock was laid down).

9 Study the following information about two clastic sedimentary rocks (A and B). For each one you are given
(i) a bar graph showing its range of grain size
(ii) a magnified view of its appearance

(a) Which of these rocks would you describe as being well sorted? (Explain your answer.)

(b) Carefully describe the texture shown by each rock (remember texture involves the size, shape and arrangement of the grains).

(c) Suggest a name for each rock type.

(d) What kinds of sedimentary environments do you think these rocks were deposited in? (Give reasons for your answers.)

10 The following sentences (a–f) describe some possible ways in which sediments are deposited. For **each** one
 (i) name the rock type that is most likely to be produced;
(ii) describe the characteristics of that rock type and explain how these are related to the conditions under which it was deposited.

(a) Clay minerals settle slowly from suspension in still waters where there is little oxygen.

(b) Large amounts of vegetation sink into the swampy waters of a tropical delta.

(c) Large rock fragments are released by frost shattering and fall to the foot of a mountain slope.

(d) Shells and shell fragments collect at the foot of a coral reef.

(e) Mixture of sand and clay-sized particles are swept down a continental slope by a turbidity current.

(f) Silt grains or tiny shell fragments are rolled across a shallow sea floor where lime 'mud' (micrite) is being deposited.

11 Explain carefully how loose deposits of sediment may be formed into solid sedimentary rocks. Your answer must include the following terms: compaction, matrix, cement, lithifaction.

12 With the aid of diagrams explain the origin of these sedimentary structures
(a) ripple marks
(b) dessication cracks
(c) sole marks

13 Carefully describe a sequence of sedimentary rocks which you have personally studied in the field. What conclusions were you able to make about the environment in which that sequence was formed?

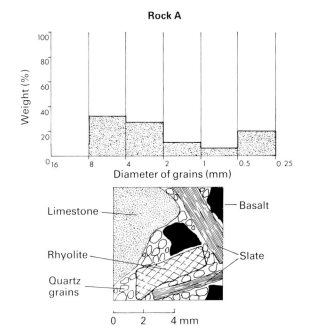

Rock A

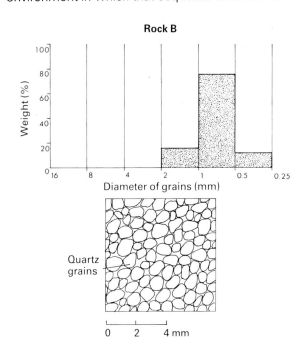

Rock B

8 Metamorphic rocks

Before beginning this chapter re-read page 27 to remind yourself of the main facts about metamorphic rocks. Remember that these are rocks which have been **changed** by **heat** and/or **pressure**. Remember, too, that metamorphic changes take place while the rock remains essentially solid. If melting occurs, then a magma is produced. Any rocks formed from this would be classified as igneous.

Let us start our study by seeing where heat and pressure are likely to occur.

Heat
- Rocks close to an **intrusion** of magma will become heated.
- Rocks **buried deep** in the crust will become heated (on average, temperature increases by about 30 °C for each kilometre of depth).

Pressure
- Rocks within the **active zones** of the earth, where plates are moving together, are subject to tremendous pressure. This is particularly true where continental edges are being deformed and uplifted into **mountain chains**.
- Rocks **buried deep** in the crust will be affected by pressure. For example, at a depth of 3 km, each square centimetre of rock will have to bear the weight of over a tonne of overlying material.

Fig. 8.1 Metamorphism: what can the earth cook up?

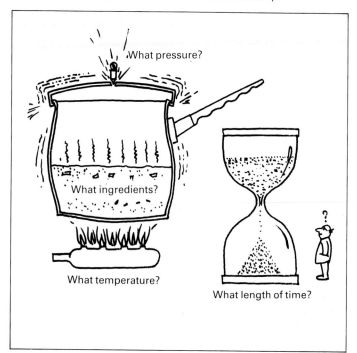

What pressure?

What ingredients?

What temperature?

What length of time?

Heat and pressure cause metamorphic change in two ways.
1. The **mineralogy** is changed: although no elements are added to or taken from the original rock, its minerals will **react** and **recrystallise** to suit the new conditions.
2. The **texture** is changed: as new minerals develop, their size, shape and arrangement will depend on the conditions under which they form.

Although heat and pressure are mainly responsible for the mineralogy and texture of metamorphic rocks, other factors need to be considered. The mineralogy of the original rock type is important because it controls what new minerals can be formed. For example, it would be impossible for calcite to crystallise during the metamorphism of a rock whose original minerals did not contain any calcium. Time is another factor in metamorphism; changes cannot take place without sufficient time for the reactions and recrystallisation to occur.

Like all other rock types, metamorphic rocks can only be described and named by studying their mineralogy and texture. An extra point of interest is trying to work out what the original rock could have been like and how much change has occurred. The word **grade** is often used by geologists when describing the amount of metamorphic change. High-grade metamorphism means that the rock has been greatly changed and looks very different from its original form. Medium grade or low grade metamorphism would mean that less change has taken place.

Metamorphic rocks are divided into two main groups according to the conditions under which they formed: (i) **thermal or contact metamorphism** and (ii) **regional metamorphism**. A third, rather less common group is produced by **dislocation metamorphism**.

Thermal or contact metamorphism

This type of metamorphism is associated with igneous activity. Magma gives out heat (**thermal energy**) which causes metamorphism of the rocks it comes into **contact** with. Pressure plays only a minor role in metamorphism of this kind.

Most thermal/contact metamorphism occurs around igneous **intrusions**. The amount of change depends largely on the amount of heat. Large batholiths cause more metamorphism than smaller dykes and sills.

The zone of changed rock around a batholith is called a **metamorphic aureole** (Fig. 8.2). Its size depends on the amount of magma and how quickly the heat was lost. Rocks are such good insulators that temperatures of several hundred degrees centigrade can exist around a deep batholith for millions of years. Nearer the surface, heat is more easily lost so metamorphism has less time to be completed.

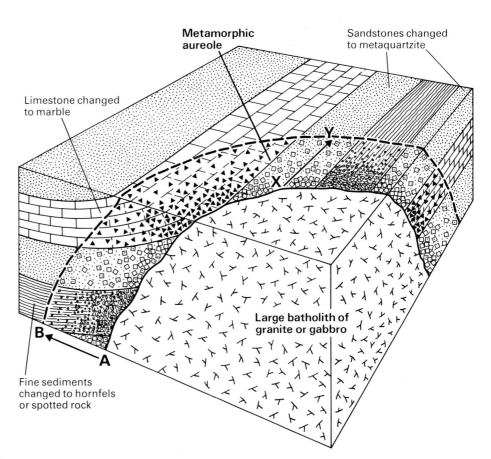

Fig. 8.2 Block diagram of metamorphic aureole in a series of sedimentary rocks. *Note*: metamorphic grade decreases from A to B and from X to Y

Variations occur within an aureole because rocks closer to the intrusion will be changed more than those further away. Sometimes definite zones of different metamorphic **grade** can be recognised (see the notes and diagrams of **spotted rock** and **hornfels** on page 83). Good examples of graded aureoles can be found around the batholiths of Scotland (for example, Insh Gabbro) and the Lake District (for example, Skiddaw Granite).

Around smaller intrusions, or beneath lava flows, there is not enough heat to produce an aureole so the changed zone is appropriately called the **baked margin** (Fig. 8.3). (Against it you will find the rapidly cooled edge of the igneous rock called the **chilled margin**.) Baked margins are often less than a metre across so it is rare to get properly developed metamorphic rock types.

A final term to remember: the original rocks into which a magma has intruded are called **country rocks**.

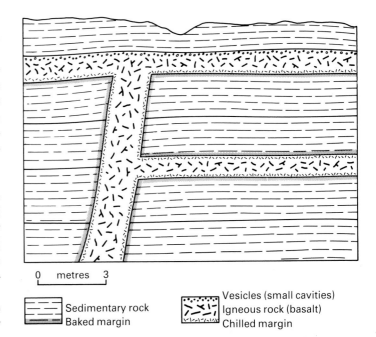

Fig. 8.3 Geological problem: baked and chilled margins. The diagram shows rocks exposed in a vertical quarry face. *Using evidence from the margins decide which igneous feature is a sill, which a lava flow and which a dyke. Carefully explain your answer.*

Regional metamorphism

As the name suggests this involves a whole region of rocks being metamorphosed. It occurs when a large volume of continental crust becomes deformed into a mountain chain by the processes of plate movement. The formation of a mountain chain is called an **orogeny** and regional metamorphism is a vital part of such an event.

The most intense changes take place at depth, so studying modern mountain chains such as the Alps or Himalayas does not give much information about the underlying metamorphic processes. Only in the up-lifted and eroded remains of ancient mountains can the full effects be seen. For example, in north west Scotland the 'roots' of mountains which formed over 1000 million years ago are exposed, and regionally metamorphosed rocks can be studied in the field.

When a whole region of rock is altered, some parts are bound to be affected by more pressure and/or heat than others. Because of this the amount (or grade) of metamorphism will vary across the area. In north west Scotland several zones of different grade can be recognised.

The most important fact about regional metamorphism is that it involves high pressure. As a result, minerals tend to become aligned, and regionally metamorphosed rocks often show a **foliated** (banded) texture. Examples include **schist** and **gneiss** which are described on page 85.

Dislocation metamorphism

Earth forces may cause rocks to fracture and move along **faults** or **thrusts** (see page 100). Any rocks close to the plane of movement are likely to be smashed or ground into fragments. A zone of broken (and therefore changed) rocks along a fault plane is called a **fault breccia**. This dislocation metamorphism may be aided by the frictional heat generated by fault movement.

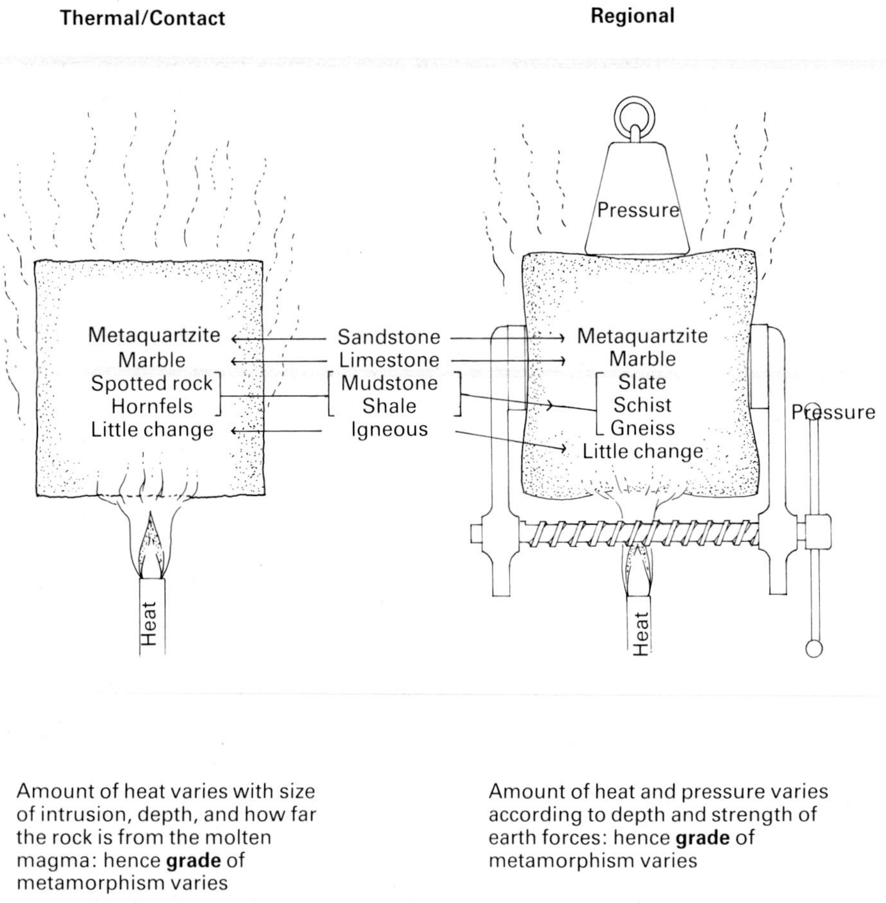

Identifying metamorphic rocks

In simple terms, Fig. 8.4 shows the rock types produced by both thermal and regional metamorphism. Since igneous rocks are formed at high temperatures, thermal metamorphism has little effect on them. Some types, particularly fine-grained volcanics, may recrystallise under the pressure of regional metamorphism. Sediments are the most likely rocks to be metamorphosed.

To understand and recognise metamorphic rocks you must study them in a practical way. The best method is to compare an original rock type with its metamorphic equivalent. For example, compare various specimens of limestone with various specimens of marble. As you do this, try to set out your results so that the difference (that is, the amount of change) is clearly shown (see Fig. 8.5). (*Note*: you can compare density by the method shown on page 23.)

The following sections give details of the main changes you should look for in your study.

Fig. 8.4 Revision diagram of metamorphism

Thermal/Contact

Regional

Metaquartzite ←	Sandstone
Marble ←	Limestone
Spotted rock ⎤	⎡ Mudstone
Hornfels ⎦	Shale
Little change ←	Igneous

→ Metaquartzite
Marble
⎡ Slate
Schist
⎣ Gneiss
→ Little change

Amount of heat varies with size of intrusion, depth, and how far the rock is from the molten magma: hence **grade** of metamorphism varies

Amount of heat and pressure varies according to depth and strength of earth forces: hence **grade** of metamorphism varies

Practical Comparision	
Sandstone	**Metaquartzite**
Mineralogy Mainly quartz Identified by hardness View x 15 Hand Lens. *Texture* Bedding clearly visible, separate rounded grains held together by natural cement (probably iron compound due to red brown colour.) *General Notes* Porous and soft. Individual grains can be broken away. Density 2·2 g/cm³	*Mineralogy* Mainly quartz. View x 15 Hand Lens. *Texture* Interlocking mosaic of quartz crystals, no spaces in between. No indication of bedding. *General Notes* Impermeable and very hard. Broken edges feel sharp. Density 2·7 g/cm³ .

Metamorphism of sandstones

(Fig. 8.6a, page 52)

Although sandstones vary in type, they always contain a high proportion of **quartz** grains. Under the effect of both thermal and regional metamorphism these grains tend to recrystallise and interlock as a solid 'mosaic' of quartz crystals. The resulting rock is extremely hard, resistant to erosion and called **metaquartzite**.

Sometimes the colour of metaquartzite may indicate how the original sandstone was cemented. For example, a reddish metaquartzite is likely to have formed from a sandstone with a cement of iron minerals.

Metamorphism of limestones

(Fig. 8.6b, page 52)

Limestones form in a variety of ways but always contain **calcite** as their main mineral. Both thermal and regional metamorphism cause this calcite to recrystallise and the rock type **marble** is produced. Pure limestones change into brilliant white marbles with a 'sugary' appearance of interlocking calcite crystals. Muddier types of limestone tend to produce marbles with patches or streaks of coloured minerals from the recrystallisation of their clay minerals.

Good examples of marble can be seen on gravestones or the fronts of important buildings such as city centre banks. However, many 'builders' marbles' are in fact just types of limestone which happen to polish well. In these you will often see the remains of fossil shells or corals. A true marble should not show such features because metamorphism would have destroyed them.

Metamorphism of mudstones and shales

Mudstones and shales are fine-grained sediments composed of minute **clay minerals**. These minerals are complex silicates containing a variety of common elements such as aluminium, magnesium, iron, calcium, sodium and potassium. Different conditions of temperature and pressure cause these elements to combine in various ways and proportions. As a result, a whole range of new minerals and rock types can be produced.

Fig. 8.7 shows what happens to mudstones and shales during **contact metamorphism**. Close to the intrusion there is normally enough heat to allow complete recrystallisation of the clay minerals, so the resulting rock has an interlocking texture and is called **hornfels**. Further away only limited recrystallisation can take place and **spotted rocks** are produced. The 'spots' are new minerals which have been able to form. Under a microscope some may appear oddly shaped or incomplete because there was not enough heat for them to crystallise properly. Within an aureole, spots of particular minerals can be used to show metamorphic grade. For example, spots of **andalusite** (Al_2SiO_5) indicate higher temperature (grade) than spots of **biotite**.

Several different grades of metamorphic rocks can be produced when mudstones and shales are affected by **regional metamorphism**. One feature shown by all these varieties is an alignment of their minerals because they all developed under the influence of pressure. Fig. 8.8 on page 52 shows examples of the rocks described below.

Low-grade regional metamorphism involves high pressure but relatively low temperatures. Under these conditions mudstones and shales are changed into **slate**. Slates were widely used for roofing because they are not only waterproof and tough but can be cleaved (split) into thin sheets. The development of this **slaty cleavage** is illustrated in Fig. 8.9. It results from the recrystallisation of clay minerals into tiny mica flakes (too small to see without a micro-

scope). Because each mica flake is a flat tabular crystal it grows to align itself at right angles to the direction of pressure. The rock can then be cleaved along these parallel mica-rich layers. Note that it is also possible for some fine-grained volcanic ashes to form slate when affected by high pressure. Some of the green slates of the Lake District are of this type.

Medium-grade regional metamorphism occurs during the development of mountain chains and involves high pressure plus moderately high temperatures. Under these conditions mudstones and shales reform into **schist**. All traces of sedimentary bedding are lost and the recrystallisation of clay minerals to micas becomes obvious enough for schists clearly to show parallel bands of **muscovite** or **biotite**. This parallel texture is called **foliation** and is a direct result of minerals growing under pressure. Sometimes other metamorphic minerals such as **garnet** (a complex silicate) develop during the formation of schists. Fig. 8.10 on page 52 shows a microscope view of a schist with micas and garnet.

Fig. 8.7 Section across a zoned metamorphic aureole

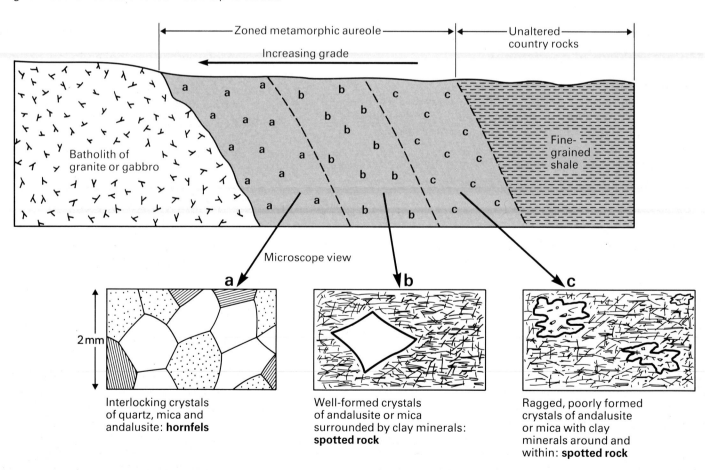

Interlocking crystals of quartz, mica and andalusite: **hornfels**

Well-formed crystals of andalusite or mica surrounded by clay minerals: **spotted rock**

Ragged, poorly formed crystals of andalusite or mica with clay minerals around and within: **spotted rock**

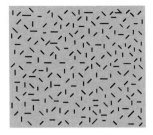

Before: mass of minute clay minerals with only limited alignment

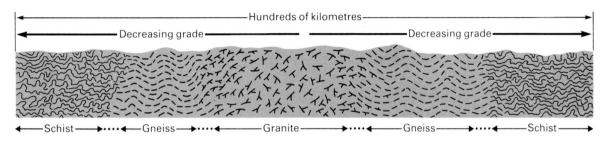

PRESSURE

Slate cleaves along parallel layers of mica flakes

PRESSURE

After: tiny mica flakes form and align themselves at right angles to the pressure

Fig. 8.9 Metamorphism of mudstone into slate

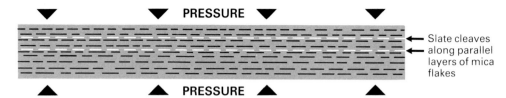

Hundreds of kilometres

Decreasing grade — Decreasing grade

Schist — Gneiss — Granite — Gneiss — Schist

Fig. 8.11 Section across the 'roots' of a deeply eroded mountain chain, showing intense regional metamorphism. In the centre, rocks have melted 'in situ' to form a granite (magma has not intruded from elsewhere). The granite merges outwards into gneiss which in turn merges into schist

High-grade regional metamorphism takes place at depth during the formation of mountain chains. The intense pressure and heat produce complete transformation of fine sediments into **gneiss** (pronounced 'nice'). As total recrystallisation takes place the rock material separates into alternate bands of light and dark minerals. The dark layers are rich in **micas, amphiboles** and **pyroxenes** while the lighter bands contain **quartz** and **feldspar**. Since the grain size is coarse it is usually possible to identify these minerals in hand specimens.

At the depths where gneisses are formed it is likely that rocks are approaching their melting point. As Fig. 8.11 shows, gneisses may merge into large masses of granite where such melting took place. In cases like these the granites have not really intruded. They have melted and crystallised at depth without the magma having moved upwards at all.

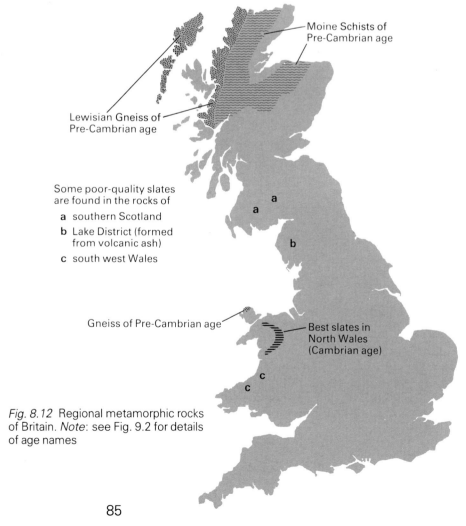

Moine Schists of Pre-Cambrian age

Lewisian Gneiss of Pre-Cambrian age

Some poor-quality slates are found in the rocks of

a southern Scotland

b Lake District (formed from volcanic ash)

c south west Wales

Gneiss of Pre-Cambrian age

Best slates in North Wales (Cambrian age)

Fig. 8.12 Regional metamorphic rocks of Britain. *Note:* see Fig. 9.2 for details of age names

Questions

For questions 1–7 write the letters (A, B, C, . . .) of the correct answers in your notebook.

1 Which **one** of the following is **not** a metamorphic change?
A change due to pressure at depth
B change due to the effects of the weather
C change due to the heat of an intrusion
D change due to the pressure of mountain building
E change due to friction caused by fault movement

2 Which **two** of the following rocks can be formed by either thermal or regional metamorphism?
A slate   B metaquartzite   C schist
D marble   E hornfels

3 Which **one** of the following rock types is formed by high-grade regional metamorphism?
A gneiss   B breccia   C slate
D spotted rock   E gabbro

4 Which **one** of the following minerals would be the most obvious in a specimen of schist?
A quartz   B feldspar   C calcite
D andalusite   E mica

5 Which **two** of the following rocks always show foliation?
A schist   B marble   C hornfels
D siltstone   E gneiss

6 Which **one** of the following rocks shows alternate bands of light and dark minerals?
A basalt   B gneiss   C metaquartzite
D slate   E spotted rock

7 Which **two** of the following minerals would you expect to find in the lighter bands of the rock described in question 6?
A amphibole   B feldspar   C biotite mica
D pyroxene   E quartz

8 The diagram below shows a section through a series of sedimentary rocks which have been metamorphosed by the heat of an intrusion.

(a) What is zone Z called?
(b) What is the correct name for this type of metamorphism?
(c) Name the igneous rock type which forms the intrusion.
(d) Name the metamorphic rock types you would expect to find at A, B, C and D.
(e) State one simple test which would easily allow you to distinguish between specimens of rock from A and D.
(f) Describe the appearance and textures that you would expect to see in specimens from B and C. How would these rocks differ from the surrounding shale?
(g) What one overall name could be given to the sedimentary rocks into which this igneous rock has intruded?
(h) Give one economic use for the rock type found at D.

9 Use the following list to choose and write down the feature most likely to be seen at the places (a)-(f) described below.
aureole / dislocation metamorphism / chilled margin / slaty cleavage / high-grade regional metamorphism / baked margin
(a) in the North Wales quarries where roofing materials were extracted
(b) in the rocks immediately above and below the Whin Sill of north east England
(c) along the line of the Great Glen Fault in Scotland
(d) in the zone of rocks surrounding the Skiddaw granite of the Lake District
(e) in the ancient rocks of the north west Highlands of Scotland
(f) in the basalt right at the edges of a dyke in the Isle of Mull, Scotland.

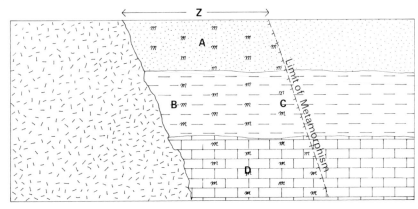

⋅⋅⋅	Sandstone
- - -	Shale
m m	Metamorphosed rock
⋰⋰	Dark coarse-grained igneous rock
▭▭	Light grey sedimentary rock composed of CaCO₃

10 Study the graph below which shows the effect of increasing pressure and temperature on fine grained sedimentary rocks.

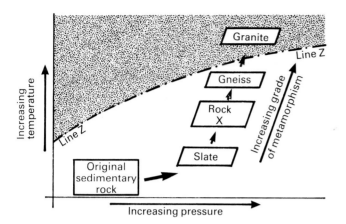

(a) Name one type of fine-grained sedimentary rock which might be affected in the way shown by the graph.

(b) Name rock X.

(c) What does line Z represent (is this really the limit of metamorphism?)

(d) Describe the changes in mineralogy and texture which occur when increasing temperature and pressure cause slate to develop into gneiss.

(e) What type of geological environment is most likely to cause rises in temperature and pressure?

(f) Is a granite magma formed in the way shown by the graph likely to move upwards and intrude into other rocks? (Explain your answer.)

11 Which one of the following rock types is least likely to be affected by thermal metamorphism and which one is likely to be most affected: basalt, mudstone, breccia, slate, gabbro? Explain your answer.

12 An area of country is described as 'being composed of regionally metamorphosed rocks'. What does this statement tell you about the geological history of the area? What rocks might you expect to find in such a place and how would you set about identifying them?

13 Explain the changes which would take place in a bed of limestone if it were
(a) intruded by a sill
(b) deeply buried during the formation of a mountain chain
(c) exposed at the earth's surface.

9 Geological time

Fig. 9.1 Some important events in the earth's history – compressed into just one year!

Fig. 9.2 The geological column

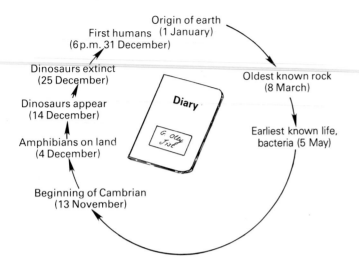

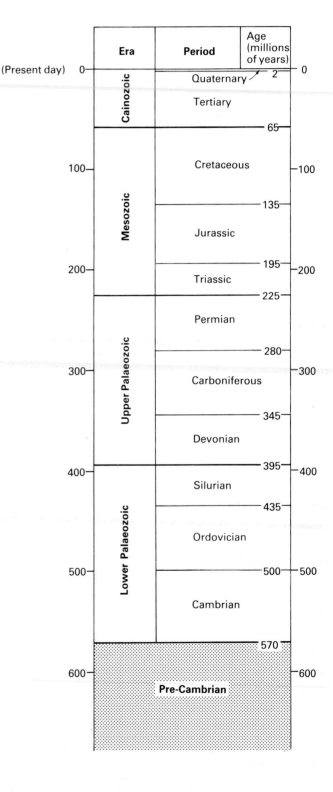

All the time in the world

Page 7 stated that the earth was formed 4600 million years ago. This figure is easy to say but can you imagine what it really means? To you, a period of ten years may seem a long time yet, as a geologist, you must deal with 'all the time in the world'. How can you open your mind and come to terms with such an idea?

Begin by looking at Fig. 9.1 which compares the earth's history (4600 million years) with one year. This will give you an idea of the time scale of different events. Notice how much time had passed before any life evolved, how long it took before creatures could live on land and, in particular, notice that humans have only appeared in the last tiny fraction of the earth's history.

Now look at Fig. 9.2. This shows the special names geologists use for certain times in the past. These names make things easier. For example, it is more convenient to describe a rock as being of Devonian age than to say it was formed some time between 345 and 395 million years ago. Each past time from Cambrian to Quaternary is a **period**. They are arranged in a column to represent the oldest rocks being buried 'at the bottom of the pile' as younger material is laid down above it.

Sedimentary rocks from each period can often be recognised by the particular fossils they contain. For example, certain fish remains are only found in rocks of Devonian age. Some types of dinosaur only lived in the Cretaceous period. Each period has a definite group of fossils which are unique to that time.

Using fossil evidence it is also possible to sub-divide the periods. The simplest method is to call the earlier part of a period the lower, and the later part the upper. For example,

Carboniferous
{
.... 280 million years
Upper Carboniferous
.... 325 million years
Lower Carboniferous
.... 345 million years
}

Larger units of time (written sideways on Fig. 9.2) are called **eras**. They are broad ages in the evolution of life. Sometimes geologists talk about eras to avoid having to list several periods. For example, it is easier to say an area contains Mesozoic sediments than to say it contains sediments formed during the Triassic, Jurassic and Cretaceous periods.

The Pre-Cambrian

You will have noticed that Fig. 9.2 divides only the last 570 million years into named units. This is unfortunately as far as the 'period system' can go because it depends on fossil evidence. Fossils are so rare in rocks older than Cambrian age that it is extremely difficult to divide the column any further. As a result, more than 4000 million years (87% of earth time) are taken together and called the Pre-Cambrian.

The Pre-Cambrian is particularly interesting. During this time the crust must have cooled and solidified, the oceans and atmosphere developed, rock-forming processes begun and, most fascinating of all, the first life forms appeared. As Fig. 9.4 shows geologists have collected evidence which allows them to date some of these events, but there is much still to learn.

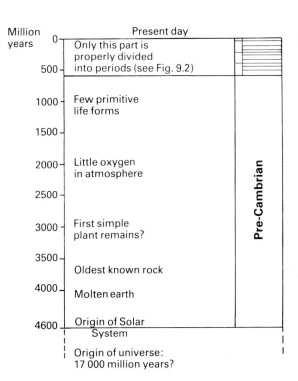

Fig. 9.4 The extent of Pre-Cambrian time

The main problems associated with investigating the Pre-Cambrian are:
- most of the rocks formed during this time have by now been destroyed by erosion or buried beneath younger rocks;
- the Pre-Cambrian rocks which do occur at the surface have nearly all been affected during their long history by metamorphism. This has destroyed or changed their original minerals and structures.

Trying to interpret Pre-Cambrian events is like trying to solve a crime after 99% of the clues have been lost, hidden or altered!

Fig. 9.3 Wenlock limestone: invertebrate fossils typical of the Silurian period

Fig. 9.5 Pre-Cambrian rocks of Anglesey: folding and metamorphism have made them very difficult to interpret

Dating the past

When studying rocks it is useful to know when they were formed. Geologists date rocks in two main ways: relative dating and radiometric dating.

Relative dating

This does *not* give an age in millions of years but allows you to work out the order in which geological events took place. For example, if a dolerite dyke is found cutting through a series of sedimentary rocks then the dyke must be younger than the sediments. The section on 'geological history' (page 110) explains the techniques of relative dating.

When the geological column was first worked out (nearly 150 years ago) it was done entirely by relative dating. Rock and fossil types were simply arranged in the order in which they had been formed, without anyone knowing exactly how old each period was.

Radiometric dating

This gives the age (in millions of years) for a particular specimen of rock. Since it relies on accurate measurement of minute traces of radioactive atoms, it is a task for specialists. All we need to know is the basic theory.

Atoms are made of tiny particles called protons, neutrons and electrons. Radioactive atoms are unstable and decay (break down) by giving out some of their particles. As a result of this decay, their internal structure changes and they become stable. (*Note*: the decay also releases energy, which is the principle of atomic power.)

Fig. 9.6 Radiometric decay and half-life (see text for explanation)

Common rock-forming minerals such as **feldspar** and **mica** contain very minute traces of radioactive elements such as rubidium (Rb) and a special form of potassium known as K^{40}. These unstable atoms are called **parent atoms** and they begin to decay into stable (**daughter**) atoms from the moment feldspar or mica are formed by crystallisation. Because the decay takes place at a standard rate (depending on which radioactive element is present), geologists can use it as an 'atomic clock' to calculate the age of the minerals.

The rate of decay is measured in terms of a unit called a **half-life**. This is the time taken for half the parent atoms to decay into daughter atoms. Fig. 9.6 illustrates the idea behind the measurements.

Diagram A shows 200 white squares each representing a parent atom (i.e. no decay has yet taken place).

Diagram B shows the situation after one half-life has passed: half the parents have now decayed and become daughter atoms (shown in black).

After another half-life (diagram C), half of the remaining parents have also decayed so only 50 (one-quarter of the original 200) white squares remain and there are now 150 black (daughter) squares.

Draw a diagram to show the situation after another half-life.

If we can measure the proportion of parent and daughter atoms in a mineral specimen, and we know the half-life of the radioactive element involved, we can work out the age of the mineral. Try these examples.

Mineral Y. There are only one-quarter of the parent atoms left (three-quarters now daughters); the radioactive element in the mineral decays with a half-life of 200 million years. *How old is this specimen?*

Number of half-lives	1	2	3	4	5
Proportion of parent atoms	$\frac{1}{2}$	$\frac{1}{4}$	$\frac{1}{8}$	$\frac{1}{16}$	$\frac{1}{32}$
Proportion of daughter atoms	$\frac{1}{2}$	$\frac{3}{4}$	$\frac{7}{8}$	$\frac{15}{16}$	$\frac{31}{32}$

Mineral Z. Fifteen-sixteenths are now daughter atoms; the radioactive element in the mineral decays with a half-life of 75 million years. *During which period was this specimen formed?*

The figures used in these examples are rather small; half-lives for the geologically useful Rb and K^{40} are each over 1000 million years.

One difficulty with radiometric dating is making sure that the mineral tested is the same age as the rock in which it is found. With igneous rocks there is usually no problem since the minerals crystallised as the rock itself was formed. However, in sedimentary rocks it is possible to date a mineral which is actually much older than the time at which the sediment was deposited. (Remember that sediments contain 'second-hand' material eroded from older rocks.)

What problems do you think might arise when dating metamorphic rocks?

Finally a mention should be made of **radio-carbon dating** which is used when dating organic remains such as plant material and bones. Unfortunately, since radioactive forms of carbon have a short half-life, this method can only be used for materials less than 50 000 years old.

Fig. 9.7 Mount Everest: 15 million years of uplift. The summit is made of rocks originally formed on the ocean floor

The speed of geological activity

How will the geology of this planet change during your lifetime? No doubt volcanoes will erupt, earthquakes occur, cliffs be cut back and sediments be deposited. In fact all the geological processes will continue in their never-ending way. However, it is unlikely that you will see any really significant geological change. A human life is too short to appreciate the rate at which geology 'happens'.

Let us illustrate this with some examples.

Rates of sedimentation. A series of Devonian age rocks in the Orkney Islands is known to be 5 km thick. Dating methods show that it was deposited over a period of 10 million years. What was the average rate of deposition?

$$\text{Total thickness} = 5 \text{ km} = 5000 \text{ m}$$
$$= 5\,000\,000 \text{ mm}$$
$$\text{Rate of deposition} = \frac{5\,000\,000}{10\,000\,000} \text{ mm/year}$$
$$= 0.5 \text{ mm/year}$$

This figure is small, but compared with many other known examples it is actually quite a rapid rate of deposition!

Rates of plate movement. The North Atlantic Ocean is 4200 km wide. It is known that Europe and North America began 'drifting' apart 60 million years ago. *What is the average rate of movement?*

A 7 mm/yr   B 70 cm/yr   C 7 cm/yr

Rates of erosion. These are very difficult to measure but we can imagine some possible effects. *Ben Nevis is 1342 m high. If 2 mm per year were eroded from its summit, how long would it take to reduce it to sea level?*

A 2684 thousand years   B 671 thousand years
C 671 million years

Note: this calculation does not allow for isostasy (see page 17). *What difference would this make?*

Remember, geological processes may seem slow but they have 'all the time in the world'.

Questions

1 Name these geological times.
   (a) the first period when fossils occur in sufficient numbers to be useful
   (b) the period which came immediately before the Cretaceous
   (c) the era which contains the Devonian, Carboniferous and Permian periods
   (d) the oldest of the time periods which make up the Mesozoic
   (e) the vast unit of geological time which ended 570 million years ago
   (f) the period which came immediately after the Ordovician
   (g) the period which lasted from 345 to 280 million years ago

2 Study the graph below which shows the rate of radioactive decay of a certain substance.

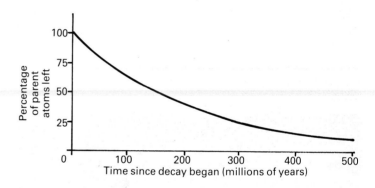

   (a) What percentage of parent atoms are left after 300 million years of decay have taken place?
   (b) What is the half-life of the substance shown on the graph?
   (c) A rock is found to contain 12.5% of the original parent atoms. How old is the rock?
   (d) Which of the following rock types could be most accurately dated by radiometric methods?
      A sandstone   *B* granite   *C* coal
      D marble   *E* shale

3 Arrange these events into the correct order (beginning with the oldest):
   development of simple plants / beginning of the Cambrian period / origin of the earth / development of soft-bodied animals / development of land-living animals / origin of the universe

92

10 Earth force: deformation in rocks

So far, this book has concentrated on how rocks are formed. Now it is time to see how they can be deformed. Deformation is caused by forces acting within the earth. It is mainly associated with geological events along plate margins (especially destructive margins where mountains are being built). In fact, the action of earth forces is often called **tectonic action**, from the Greek word 'tecton' meaning builder.

Tectonic action is responsible for the formation of:
(a) **folds**,
(b) **faults**,
(c) **cleavage** (in rocks),
(d) **joints**.

The style of deformation does not just depend on the type and direction of force applied. The nature of the rock is also important. For example, a hard brittle sequence of crystalline rocks will probably fracture (fault) when force is applied but a series of softer sedimentary rocks is more likely to crumple (fold). Also, force (especially at depth) may cause metamorphism as well as deformation.

Fig. 10.2 Small anticline seen in a quarry face near Ffostrasol, Dyfed

Fig. 10.1 Rock showing the effects of deformation

5 cm

Folding

Rocks become folded when they are affected by forces of compression (compression involves 'pushing together' or 'squeezing' – like the forces that act when you tighten a vice).

Fold structures are most obviously seen in sedimentary rocks. Beds which have been folded will no longer be lying in their original horizontal position; the forces of compression will have 'buckled' them so that they **dip** (tilt) at an angle (see Fig. 7.4 on page 65).

Although a range of fold structures can occur they are all really variations of two main types: the upward fold (an **anticline**) and the downward fold (a **syncline**). Figs. 10.2 and 10.3 show examples of these folds

Fig. 10.3 Syncline seen in a sea stack at Stackpole, Dyfed

while Fig. 10.4 illustrates how both types may be found together in a fold series.

To demonstrate the development of fold structures you can easily make a 'working model' of Fig. 10.4. Use sheets of paper or flat layers of Plasticine for the sedimentary beds and then deform your model by pushing its edges together. *What is the relationship between the direction in which you push and the direction in which the folds develop? Is there any relative movement between individual layers (beds) as folding takes place; for example, are they forced to slip against each other (along the bedding planes) as they move into their folded positions?*

In order to interpret fold structures, a good geologist must first be able to describe and measure them accurately. Make sure you use the correct terms (given in Fig. 10.5) when describing the folds you see. Some folds will need several terms to describe them properly. *Try to draw a rounded asymmetrical syncline and a tight symmetrical angular anticline.*

You will see from Fig. 10.5 that the shape of a fold depends particularly on the dip of its **limbs**. As part of your fieldwork you will need to measure the dip of fold limbs yourself; in fact, you will need to make a series of measurements known as **dip** and **strike**.

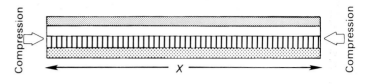

(a) Original sequence with horizontal beds

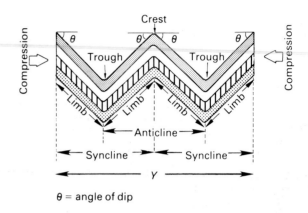

(b) Folded sequence with dipping beds

θ = angle of dip

Fig. 10.4 Folding (in sedimentary rocks). Notice (i) how compression has reduced the overall length of the sequence from distance X to distance Y: in other words folding is accompanied by crustal 'shortening'; (ii) how each fold limb may be part of both an anticline and a syncline

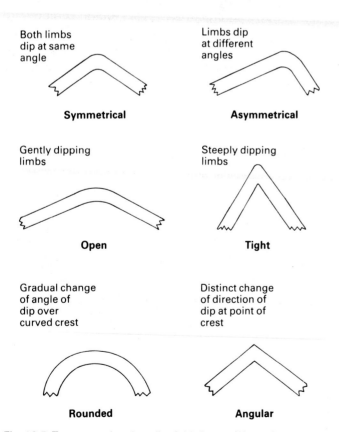

Fig. 10.5 Terms used to describe fold shapes (*Note*: these examples are all anticlines but the terms can also be used for synclines)

Dip and strike

Recording the dip and strike of a bed of sedimentary rock involves finding a suitable bedding plane and taking measurements which give the exact orientation of that plane. The only real way to learn the technique is by field practice but (as a guide) the main aspects of measurement are explained below and illustrated by Fig. 10.6.

Angle of dip
The angle of dip is measured between the **bedding plane** and the horizontal. An instrument called a **clinometer** is used. You can make one from a protractor and a piece of board but be sure to check that it reads 0° on a horizontal surface and 90° on a vertical one (like the clinometer shown in Fig. 10.6). When measuring the angle of dip you must keep the base of the clinometer firmly in contact with the bedding plane and position it so that it measures the maximum (steepest) angle of dip on that plane.

Direction of dip
The direction of dip is measured with a compass. In Fig. 10.6 the direction of dip is to the east.

Direction of strike
Strike is defined as a horizontal line drawn on the bedding plane at 90° to the direction of dip. (Since

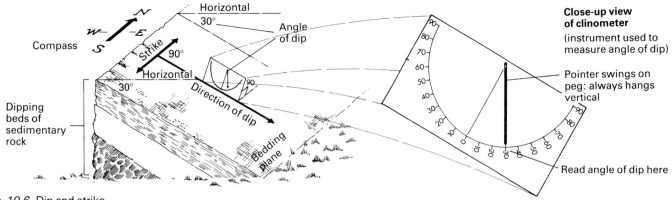

Fig. 10.6 Dip and strike

strike is a horizontal line, a clinometer placed along it will read 0°.) The direction of strike is measured with a compass. In Fig. 10.6 the direction of strike is north–south.

When measuring dip and strike in the field most geologists prefer to work in this order.
(1) Place the clinometer on the bedding plane and move it around until a reading of 0° is obtained. Draw a line on the bedding plane to mark where the base of the clinometer rests on the rock. Since this line is horizontal it will mark the direction of strike.
(2) Draw another line on the rock so that it is at right angles to the direction of strike and points directly down the sloping bedding plane. This line will mark the direction of dip.
(3) Using a clinometer, measure the angle of dip (do this by placing the clinometer on the line marking the direction of dip).
(4) Using a compass, measure the direction of strike and the direction of dip (as shown by the lines drawn on the rock).
(5) Write your results in this form:
Strike = north–south / Dip = 30° to the east.
(Note that this example uses measurements shown by the bedding plane in Fig. 10.6; when writing down measurements taken in the field you should also add the grid reference of the locality.)

Folds in three dimensions

Fig. 10.7 is a block diagram drawn to show the pattern of beds produced when the 'top' of the folds are 'cut away'. Patterns like these are often seen in folded areas which have been affected by erosion. Note the **fold axes**: these are imaginary lines drawn along the centre of each fold. In anticlines they follow the line of the **crests**, in synclines the line of the **troughs**. The direction of a fold axis is called its **trend** and in Fig. 10.7 all the axes trend north–south. A careful study will show that fold axes are at 90° to the

direction in which the beds dip (beds dip either to the east or to the west in the diagram). *What is the relationship between the fold axes and the directions of strike of the beds?*

Fig. 10.7 also gives the age order of the beds. Remember that when these rocks were originally deposited the oldest bed would have been at the bottom of the sequence and the youngest at the top. Folding and erosion can have the effect of bringing lower (older) rocks closer to the surface than they would otherwise have been. Similarly upper (younger) beds may be folded down to lower levels than they would otherwise have been. Use Fig. 10.7 to check this geological rule: the oldest beds to be seen in a folded sedimentary sequence will be along the axes of eroded anticlines; the youngest beds to be seen will be along the axes of eroded synclines. (Note that this 'rule' can be broken in rare cases where the whole sequence has been completely overturned.)

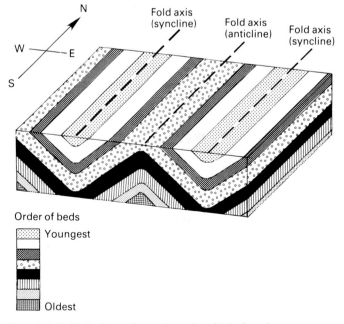

Fig. 10.7 Folds in three dimensions. *In which direction must compression have been acting to produce this direction of folding?*

Overturned beds

Where compression is really intense and/or where sediments are particularly soft, beds may be folded to such an extent that part of the sequence is overturned (see Fig. 10.8). Folds with an overturned limb are described as **overfolds**. In extreme cases where the overturned limb lies almost horizontally, the term **nappe** (pronounced 'nap') is used. *Study Fig. 10.9 which shows the increasing effects of compression; in each case decide whether limbs X, Y and Z are the right way up or overturned.*

When examining folds in the field you will have to use some geological detective work to discover if a limb has been overturned. Look for clues such as those shown in Fig. 10.11.

Fig. 10.8 Overturned beds. *Try this for yourself by laying down a sequence of flat sheets of paper then folding them over as shown.*

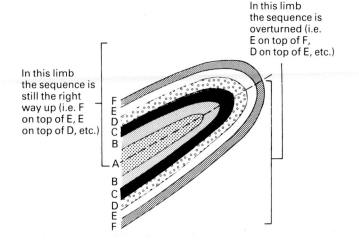

Originally sediment A was laid down first (oldest), this was then covered by B, followed by C, D and E until bed F formed the last (youngest) layer.

In this limb the sequence is overturned (i.e. E on top of F, D on top of E, etc.)

In this limb the sequence is still the right way up (i.e. F on top of E, E on top of D, etc.)

Cleavage in rocks

Slate has already been mentioned as an example of a rock which may be cleaved into thin sheets. Re-read page 84 to remind yourself how slate's cleavage is caused by **clay minerals** realigning themselves and recrystallising under the influence of pressure (compression). Since compression also causes folding, many rocks show the combined effects of both cleavage and folding. Fig. 10.12 illustrates the relationship between these features. Notice that, because the same direction of compression has been responsible for both, the cleavage planes are parallel to the fold axis.

Although slates show the best development of cleavage, other fine-grained rocks such as **mudstone** and **volcanic ash** (tuff) may also be affected to a lesser extent. When examining these rocks in the field you will often see two sets of planes (cleavage and bedding). It will be up to you to decide which is which, and only practice can teach you this skill.

Fig. 10.10 Overfold seen in cliffs at Broadhaven, Dyfed

Fig. 10.9 Development of overfolds and nappes

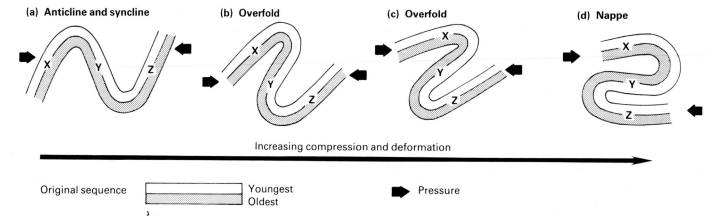

(a) Anticline and syncline   (b) Overfold   (c) Overfold   (d) Nappe

Increasing compression and deformation

Original sequence — Youngest / Oldest   Pressure

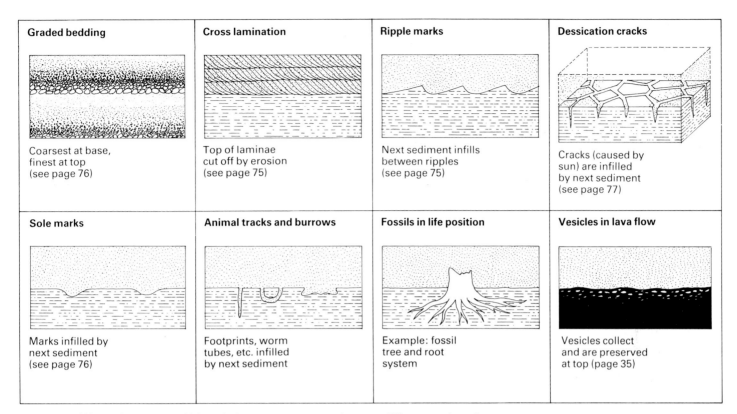

Graded bedding	Cross lamination	Ripple marks	Dessication cracks
Coarsest at base, finest at top (see page 76)	Top of laminae cut off by erosion (see page 75)	Next sediment infills between ripples (see page 75)	Cracks (caused by sun) are infilled by next sediment (see page 77)
Sole marks	Animal tracks and burrows	Fossils in life position	Vesicles in lava flow
Marks infilled by next sediment (see page 76)	Footprints, worm tubes, etc. infilled by next sediment	Example: fossil tree and root system	Vesicles collect and are preserved at top (page 35)

Fig. 10.11 'Way up' structures. Although these structures are drawn at different scales, all are shown the right way up. If you ever find them upside down you will know powerful tectonic forces have overturned the rock sequence

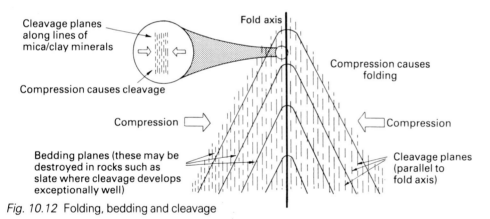

Cleavage planes along lines of mica/clay minerals

Compression causes cleavage

Fold axis

Compression causes folding

Compression

Compression

Bedding planes (these may be destroyed in rocks such as slate where cleavage develops exceptionally well)

Cleavage planes (parallel to fold axis)

Fig. 10.12 Folding, bedding and cleavage

Fig. 10.13 Development of a fault

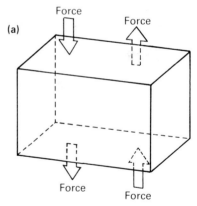

(a)

Force
Force
Force
Force

Faulting

Faulting occurs when rocks are fractured by tectonic force. A **fault** is really a break which allows rocks on one side to move against those on the other side. Fig. 10.13 gives the basic idea (much simplified) but it only shows a vertical **fault plane** and vertical movement of the rock masses. In real situations many variations are possible depending on the direction and type of force and the structure of the rocks involved.

As explained on page 10, the fracturing of rocks causes **earthquakes**. However, shock waves are not just released at the time when the break (fault) first occurs. As long as tectonic forces keep acting, rocks will tend to continue moving against each other along the line of the fault. Because of friction between the broken 'edges', such movement usually takes the form of a series of sudden jerks each of which also produces an earthquake. Many faults show the effects of millions of years of force:

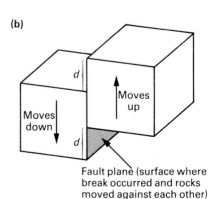

(b)

d Moves up

Moves down

d

Fault plane (surface where break occurred and rocks moved against each other)

d displacement (total distance moved)

their large **displacement** (possibly several kilometres) is the combined result of a great many small movements of just a few centimetres or metres each time.

Vertical fault movement

With this type of faulting geologists talk of an **upthrow** side and a **downthrow** side (see Fig. 10.14). However, these terms only describe the *relative* position of the rocks after faulting has occurred (that is, which side is now found relatively higher up and which side relatively lower down): they do not imply that a particular side has definitely moved up or down. For example, the rock masses A and B in Fig. 10.14 could have moved to this position in any one of the following ways.

- A moved up while B moved down
- A moved up while B remained still
- B moved down while A remained still
- A and B both moved up but A moved further than B
- A and B both moved down but B moved further than A

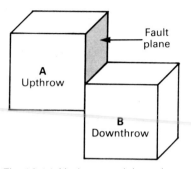

Fig. 10.14 Upthrow and downthrow

Fault planes are often sloping. In this situation two different fault types can develop: **normal faults** and **reverse faults**. (see Fig. 10.15). Normal faults are caused by forces of tension (pulling apart) which tend to increase the overall length of a particular unit of rocks. Reverse faults are caused by forces of compression (pushing together) and tend to shorten the overall length of rock units. Since both these fault types involve vertical displacement they have an upthrow and a downthrow side.

Normal faults (Fig. 10.16) are far more common than reverse faults because compression is more likely to cause folding than to produce (reverse) faulting. Normal faults may also be responsible for landscape features such as **rift valleys** and **horsts** (see Fig. 10.17).

Fig. 10.15 Normal and reverse faults

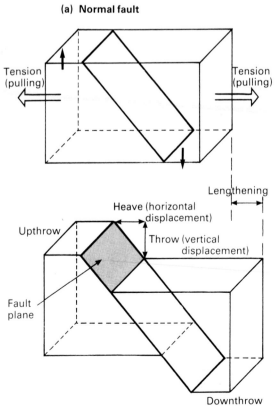

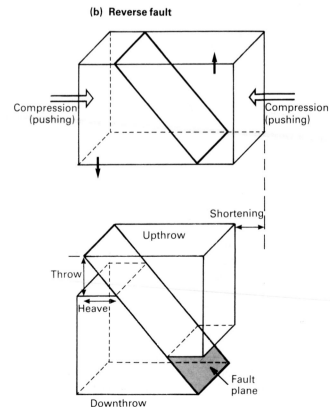

Fig. 10.16 Small normal fault seen in cliffs at Staithes, North Yorkshire

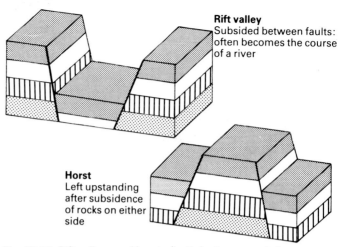

Rift valley
Subsided between faults: often becomes the course of a river

Horst
Left upstanding after subsidence of rocks on either side

Fig. 10.17 Rift valleys and horsts (both features result from movement on sets of normal faults, i.e. tension is responsible)

Horizontal fault movement

Fig 10.18 shows the effect of rock masses moving horizontally along a fault plane. These **tear** or **wrench** faults are easiest to see where one particular geological structure (for example, a dyke) has been clearly **offset** (moved out of line).

The best example of a tear fault in Britain is the **Great Glen Fault** (see Fig. 10.19). The displacement can be measured as the distance between the two halves of one granite batholith which, over millions of years, slowly slipped sideways, moving further apart.

The largest horizontal displacements occur at faults where two of the earth's plates 'slide' past each other. The **San Andreas Fault** beneath San Francisco is of this type (see Figs. 3.12 and 3.13, page 146).

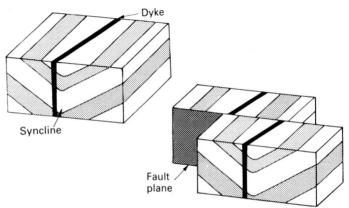

Fig. 10.18 Block diagram of a tear fault (note how geological features are offset by the horizontal movement)

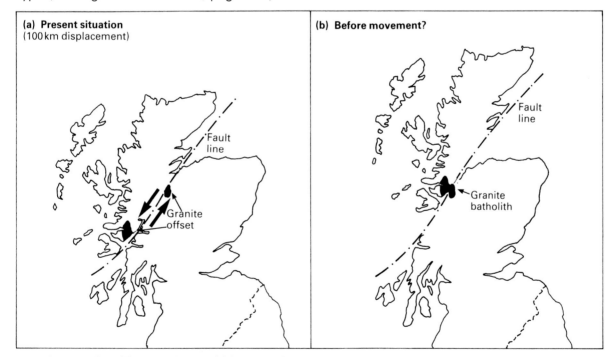

(a) Present situation
(100 km displacement)

Fault line

Granite offset

(b) Before movement?

Fault line

Granite batholith

Fig. 10.19 Great Glen tear fault (The coastline would, however, have been very different in Ordovician times before the fault began moving. Therefore, Scotland never looked exactly like map (b).)

Thrust faults

When compression is really intense, folding may reach the point where the beds are so contorted that they must fracture in order to allow any more movement. In such situations **thrust faults** develop. As Fig. 10.20 shows, this allows part of an **overfold** or **nappe** to move up and over the lower part. A thrust fault is really a special kind of reverse fault with an almost horizontal fault plane. Notice how faulting of this type allows a great deal of crustal 'shortening' to take place.

Thrusting is particularly associated with the intense forces of mountain building. Sometimes older rocks are pushed up and over younger material. At the **Moine Thrust Fault** in north west Scotland Pre-Cambrian rocks are found above Cambrian rocks (see Fig. 14.10 on page 158).

Fault plane features

The tremendous friction and stress which develop on a fault plane have geological effects of their own. Many planes contain a mass of crushed and broken rock fragments called a **fault breccia** (see section on dislocation metamorphism, page 82). Other fault planes show deep grooves called **slickensides**. The direction of the grooves indicates the direction of fault movement.

Fault planes also act as 'gaps' through which fluids can flow. In some cases the fluid is magma and a **dyke** crystallises along the fault plane. In other cases the fluids may carry minerals in solution (see section on hydrothermal solutions, page 176). Many of the veins of galena and fluorite found in the Pennines are along faults.

Erosion and faults

The position of faults is sometimes shown by landscape features. Many faults act as lines of weakness and are eroded to form valleys (Fig. 10.22b). This has happened, for example, along the line of the Great Glen Fault. Fault scarps occur where ground rises steeply to the higher level of an upthrown block (Fig. 10.22c).

Fig. 10.22 Possible effects of faulting and erosion. (a) Movement but no erosion; (b) the weakened rocks along the fault line have been eroded to form a valley; (c) a fault scarp is formed by the upthrow side standing above the downthrow side. *Note:* all the diagrams are cross-sections

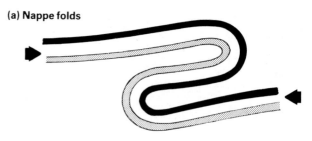

(a) Nappe folds

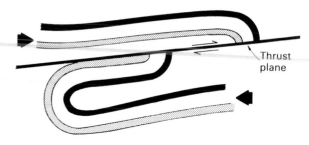

(b) Nappe folds with thrust fault

Fig. 10.20 Development of a thrust fault

Fig. 10.21 Thrust fault seen in cliffs at Broadhaven, Dyfed

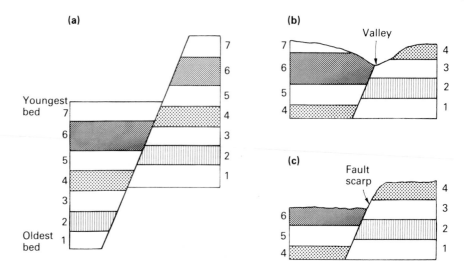

As Fig. 10.23 shows, the erosion of a faulted area can produce a pattern of rocks at the surface which (at first sight) may be misleading about the type of movement that has taken place. The best way of studying how such surface patterns can be produced is to make simple models showing beds of dipping sedimentary rock (from layers of Plasticine) then fault and 'erode' them by cutting through.

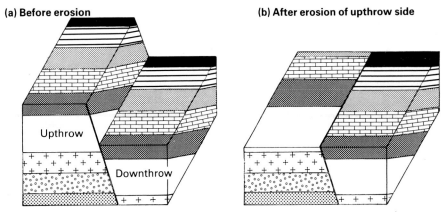

(a) Before erosion

(b) After erosion of upthrow side

Fig. 10.23 Erosion and faults. A three-dimensional diagram showing one possible effect of normal faulting on a sequence of dipping rocks. (a) Movement but no erosion (b) Erosion of upthrow side leaves a geological pattern which, from the surface, appears to have been caused by tear faulting (note the offset of beds)

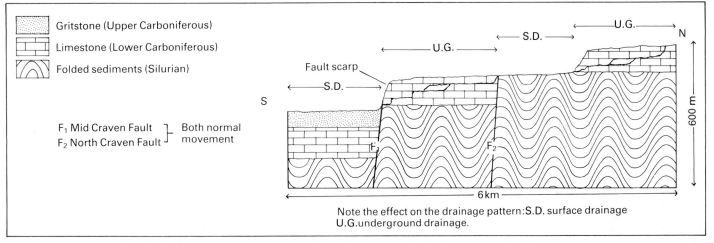

Fig. 10.24 Example of faulting in the Pennines: a north–south section through the Malham area of Yorkshire

Joints

Fractures often occur without the rocks on either side moving relative to each other. These cannot be called faults (there is no displacement) so the term **joint** is used. Joints form in several ways.

(a) Some sediments, particularly hard limestones and sandstones, develop sets of joints during folding. Fig. 10.25 shows the typical pattern. Try to produce this yourself by bending a thick pencil eraser and noting how it cracks.

(b) Igneous intrusions form at depth. By the time they are seen at the surface, a large amount of overlying rock has been eroded away. This action releases pressure and allows the upper part of the intrusion to 'crack open' and develop a set of joints.

(c) The contraction of thick lava flows (during cooling) may cause **columnar jointing** (see Fig. 5.9 on page 35).

Joints may make a rock permeable (see page 50 to revise about joints in limestone). Another of their effects is to provide 'weaknesses' which weathering can take advantage of (for example, frost shattering, page 43).

Some mineral formations occur in joint systems. For example, joints in the Shap Granite of the Lake District have been infilled by pyrite, chalcopyrite and fluorite.

Finally, joints are welcomed by quarry workers who make good use of these natural weaknesses when extracting the rock.

Look out for joints on your fieldwork and learn to distinguish them from bedding and cleavage planes.

Fig. 10.25 Joint sets in folded rock

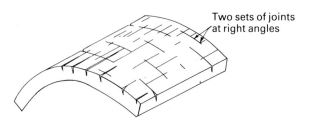

Two sets of joints at right angles

101

Unconformities

An **unconformity** is produced when the deposition of sediments is interrupted. Fig. 10.26 shows how this can be caused by uplift and erosion. Study the diagrams and notice that the unconformity represents a 'gap' in the geological sequence (a time when no deposition took place). Calling a 'gap' like this an unconformity is very appropriate because the rocks above and below will not **conform** with each other. They will have different ages and, in all probability, show different features and structures. (*Note*: not all unconformities show such a marked difference in structure as is illustrated by Fig. 10.26. It is possible, for example, to have horizontal rocks above and below an unconformity: the really important feature is that an interruption of sedimentation has caused part of the sequence to be 'missing' from this point.)

A major unconformity in British geology developed when Lower Palaeozoic rocks were folded and uplifted. By the time the sea spread back over these rocks and deposition began again (in the Lower Carboniferous), much of the original sequence had been eroded away. Fig. 10.27 shows one place where this unconformity can be seen at the surface (see also Chapter 14).

Fig. 10.27 Unconformity at Thornton Force, North Yorkshire. Here horizontal beds of limestone (Carboniferous age, about 320 million years old) are seen lying unconformably on vertical folded beds of slate (Ordovician age, about 470 million years old)

Fig. 10.26 Development of an unconformity (U labels the unconformity surface)

(a) Series of sediments deposited (usually below sea level)

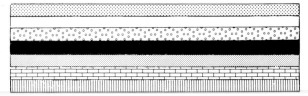

(b) Sedimentation stops as rocks are uplifted and folded

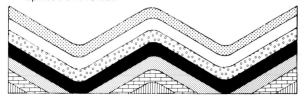

(c) While this area is uplifted no sediments are deposited so a 'gap' is produced in the geological sequence. (In fact, during this time erosion is removing part of the original series of sediments.)

(d) Sedimentation begins again, usually when the area is once again covered by the sea after a **marine transgression**. The new series of rocks is deposited **unconformably** on the eroded remains of the old.

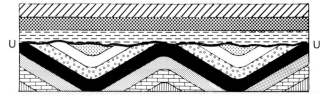

(e) Further uplift and erosion eventually reveal the unconformity at the earth's surface.

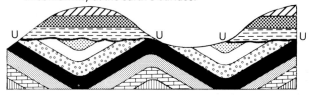

Questions

Questions on this chapter are combined with those for chapter 11 and can be found on page 112.

11 Geological mapwork

A map is an ideal way to illustrate the features of a particular area. Ordinary maps deal with the positions of things like hills, rivers, towns and roads. Geological maps show the rocks on which the area is founded. Two main types of geological maps are produced. Each can be in a variety of scales.

- **Drift maps.** These show the distribution of loose **surface deposits** such as glacial till, river alluvium, moraines, peat, etc.
- **Solid maps.** These show the distribution of the 'solid' **bedrock**. This is what you would see **exposed** at such places as cliffs and quarries or what you would meet after digging through any covering of soil or drift. Solid maps also show the position of structures such as faults, folds and unconformities.

Have a look at some geological maps of your area (the local library should have copies). Find out what bedrock and drift lie beneath your home or school. Can you see some relationship between the geology and the landscape?

To understand geological maps properly you must realise that they can only show the distribution of rocks and structures at the earth's surface. You have to interpret this surface pattern to gain a full three-dimensional picture of the geology it represents. Fig. 11.1 illustrates this point. Although the block diagram is able to show the solid geology in three dimensions, the map can only give a two-dimensional view of the surface plus some clues about the structure below. Such clues include arrows to indicate the angle and direction in which beds dip, and dashes on fault lines to show which way movement has occurred. The pages which follow explain how you can use clues like these to help you draw cross-sections that illustrate the 'hidden dimension' of geological maps. (Remember that a cross-section is the view you would get by slicing straight down through the land and looking at the cut edge.)

Fig. 11.1

(a) Three-dimensional (block) diagram showing the solid geology of a small area

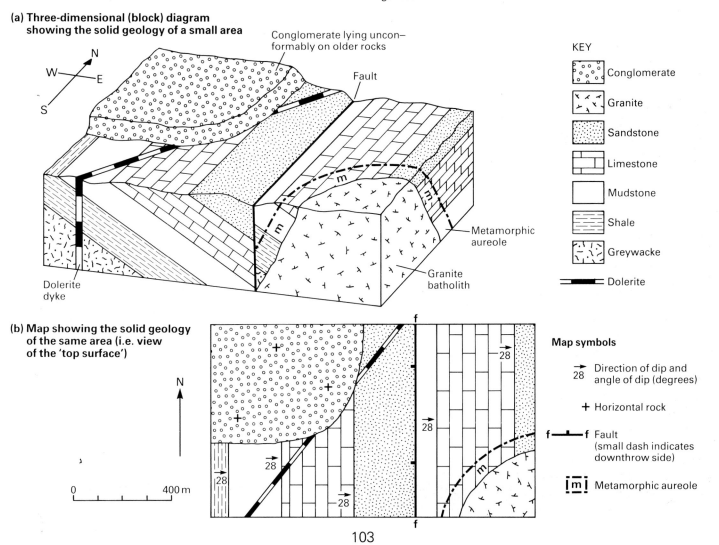

(b) Map showing the solid geology of the same area (i.e. view of the 'top surface')

103

Geological cross-sections: folded rocks

The object of this exercise is to show the third dimension of Map 1 by drawing a cross-section between point X and point Y. To help in this task, a blank section (called a **profile**) is provided. The top line of this profile represents the shape of the land surface between X and Y.

Constructing the section

Note. All section drawing must be done with a sharp pencil. The first task is to transfer information about the surface geology from the map to the top line of the profile. This is done by laying a strip of paper against a line drawn from X to Y on the map. Each rock type and each geological boundary is carefully marked onto the strip and all dips are noted (angle and direction). The ends of the section (X and Y) are also marked as reference points on the strip. An example of a properly marked strip is shown in position on Map 1.

Map 1

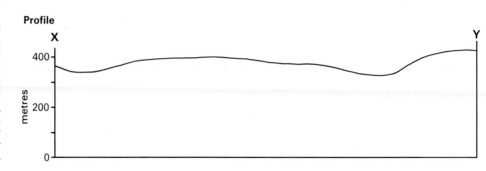

Strip of paper placed along section line

KEY

Mudstone (Mst)   Siltstone (Silt)

Shale   Sandstone (Sst)

Profile

Now place the marked paper strip exactly below the profile (see Section 1A). Make sure it is the right way round (in this example X against X and Y against Y).

From each mark on the strip project a faint pencil line vertically to the top (surface) of the section. Then use a protractor and plot the directions and angles of the dips as shown. Remember that dip is measured from the horizontal: so you may need to draw small horizontal guide lines to help. Label the rock types along the section top.

Section 1A

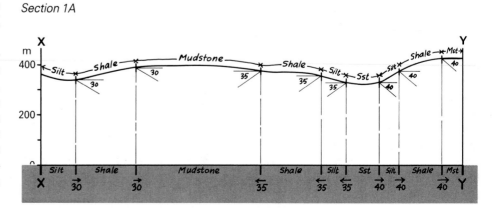

Remove the paper strip and rub out the vertical construction lines. Now extend each dipping line (geological boundary) across the section (see Section 1B). You do not need to draw them very far: stop when a line crosses another one. Note that sometimes lines need to be continued above or below the section before crossing occurs.

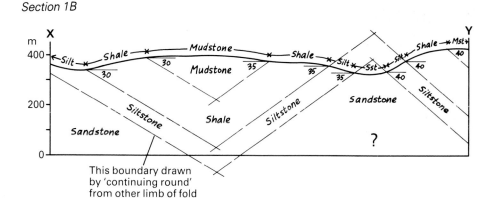

Section 1B

This boundary drawn by 'continuing round' from other limb of fold

Check that the set of lines makes a logical fold pattern. For example, if on your section a sandstone bed apparently dips below the surface and then reappears elsewhere as a shale, something must be wrong!

To complete the pattern it is often necessary to 'continue round' a particular bed. For example, in Section 1B, the sandstone/siltstone boundary has been continued round the trough of the syncline to complete the sequence in the left-hand part of the section.

Double check your fold pattern.

- Does the order of beds remain the same everywhere on the section? (For example, in Section 1B mudstone always lies above shale, shale is always above siltstone, etc.)
- Does the fold pattern join surface outcrops in a logical way? (For example, does sandstone join up with sandstone, etc.)
- Does each particular bed stay the same thickness right across the section?

If everything seems correct, neatly shade the section using exactly the same symbols as the original map key. Rub out any unnecessary construction lines or labels.

Make the folds more realistic by rounding their crests and troughs and, where possible, draw the rock symbols to follow the fold pattern (note how the shale 'dashes' are drawn in our completed example of Section 1C).

It is a good idea to mark the possible base of the lowest bed with a dotted line then add some question marks below. This shows that you realise other beds must lie beneath those you have been able to draw from the information on the map.

Section 1C

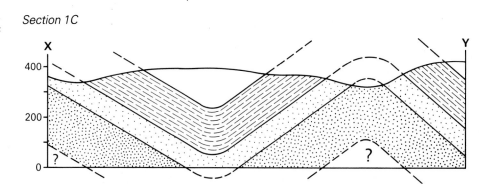

Geological cross sections: faulted rocks

Faults should present no problems when drawing cross-sections. Look at Map 2. Since both faults cut across the map as straight lines you can assume the fault planes are vertical. Only the fault (f_1) which crosses the section line needs to be plotted on the profile.

Map 2

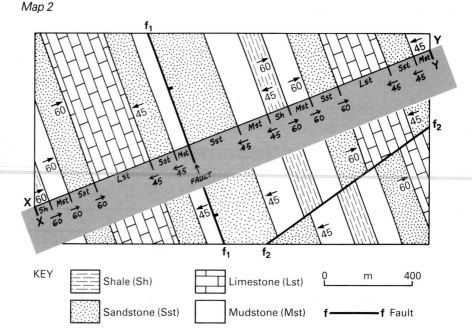

KEY

Shale (Sh) Limestone (Lst)

Sandstone (Sst) Mudstone (Mst)

0 m 400

f ——— f Fault

Use a strip of paper to collect information from the map (check the example on Map 2 and note how the fault is clearly labelled).

Put the strip against the section base. First mark the vertical fault (use a thicker line than normal) then plot rock types and dipping boundaries in the normal way (as shown in Section 2A).

As you construct the fold pattern remember that a fault is a place where rocks have been broken. With this in mind you will not forget that boundaries *stop* when they meet a fault. In effect, a fault separates (or **breaks**) your section into two distinct units.

Section 2A

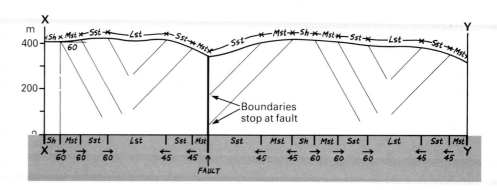

Boundaries stop at fault

Finish the section in the usual style with neat shading. Any vertical displacement (throw) should be easy to see. In Section 2B, the throw is measured by line T and, using the scale, we can work out that this equals 280 metres.

Without drawing another section, describe what type of movement has taken place on fault f_2. Try to measure its displacement.

Section 2B

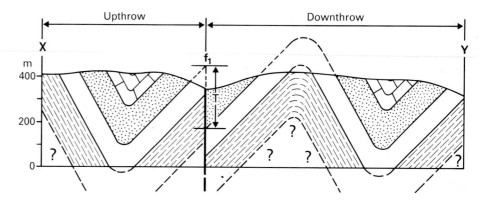

Upthrow Downthrow

Geological cross-sections: igneous rocks

Igneous rocks occur in a variety of ways (refer to Fig. 5.16, page 39). The most common features on maps are (a) lava flows, (b) beds of ash or tuff, (c) sills, (d) dykes, (e) batholiths. Features (a), (b) and (c) all occur parallel to the bedding of sedimentary rocks and present no problem in sections. Simply draw them at the dip shown. The way to plot **dykes** and **batholiths** is explained in the example below.

The section line on Map 3 crosses two dykes and a batholith (with metamorphic aureole). The surface outcrops of these features are plotted on the strip together with information about the surrounding sediments.

With the strip in position, project all marks to the top (surface) of the profile (Section 3A). Then draw the dykes as vertical intrusions: shade them straight away to remind you what they are. From the points marking the surface outcrop of the batholith, draw lines dipping steeply outwards (see Section 3A). This illustrates the typical 'up-turned boat' shape of such intrusions. Add broken lines parallel to this for the metamorphic aureole. You will know its width from marks on the paper strip.

Now plot sedimentary boundaries but note what happens when they meet an intrusion. Boundaries stop at dykes but reappear on the other side. At batholiths it is best just to stop the boundaries, although on some sections it is possible to work out how beds must have joined up before the batholith intruded into them. (For example, note how the anticline has been drawn on Section 3B.)

Map 3

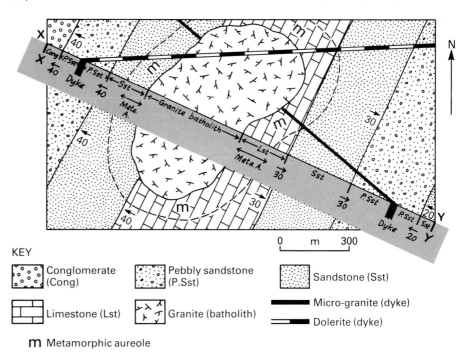

KEY

- Conglomerate (Cong)
- Pebbly sandstone (P.Sst)
- Sandstone (Sst)
- Limestone (Lst)
- Granite (batholith)
- Micro-granite (dyke)
- Dolerite (dyke)
- m Metamorphic aureole

0   m   300

Section 3A

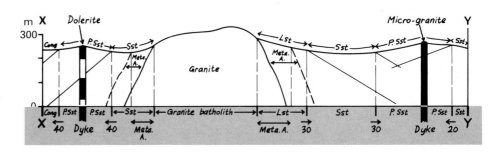

Section 3B

Geological cross-sections: unconformities

When an unconformity occurs on a section you have the extra task of deciding what structures are buried beneath it. Look at Map 4A. Here a folded eroded sequence of sediments (plus a dyke) is partly hidden by basalt lava and volcanic ash lying unconformably above.

Use a strip of paper to mark the surface outcrops as shown. Clearly label which parts are above the unconformity.

Mark from the strip to the top (surface) of the profile in the usual way. Note how the horizontal beds are plotted and that the unconformity is clearly labelled.

You now have to work out what rocks and structures lie buried beneath the unconformity. Do this by extending the mapped boundaries of the older rocks across the area now 'hidden' by the unconformity; that is, redraw the map as it would look with the newer unconformable rocks 'lifted off'. (See the example given by Map 4B.)

Take a *new* paper strip and mark the 'buried boundaries' where they cross the section line.

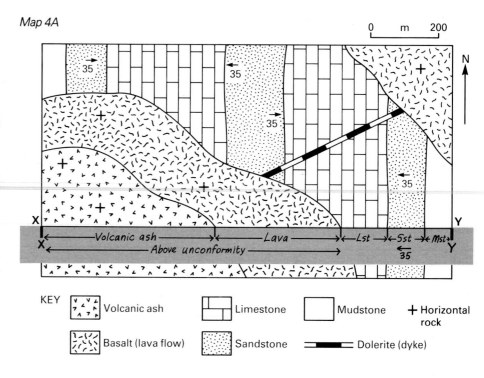

Map 4A

KEY

Volcanic ash · Limestone · Mudstone · + Horizontal rock

Basalt (lava flow) · Sandstone · Dolerite (dyke)

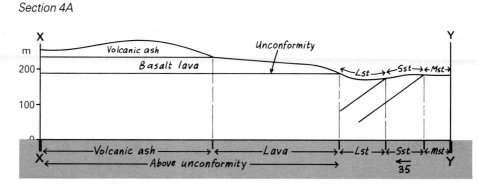

Section 4A

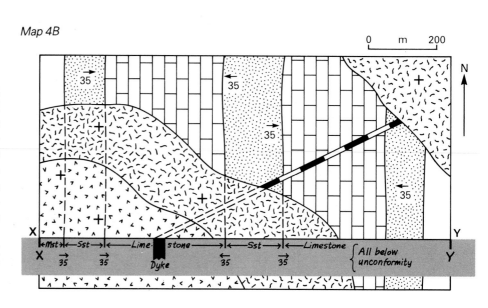

Map 4B

Put the 'buried boundary' strip against the profile and *project up to the base of the unconformity* (see Section 4B). Plot dips, etc. in the normal way.

Section 4B

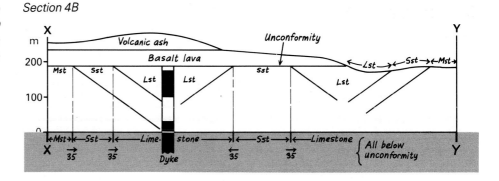

Complete the structure of the older rocks by joining boundaries (some exposed at the surface and some below the unconformity). Double check your work before shading the section neatly. (See Section 4C.)

Section 4C

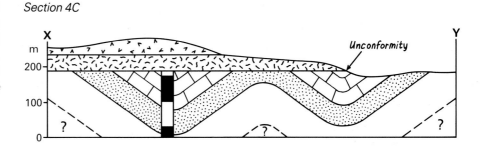

Surface deposits

Unconsolidated surface deposits (drift) are shown on certain types of map. These lie unconformably on the solid rocks beneath, so section drawing uses the method explained for Map 4. Map 5 shows a typical example.

Map 5

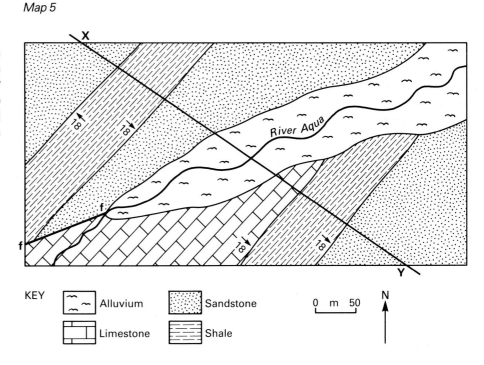

KEY   ≈≈ Alluvium   ∴ Sandstone   0 m 50   N

Limestone   Shale

Section 5

109

Interpreting the past: geological history

As you already know, geological processes have been at work for thousands of millions of years. This means that any particular area will have experienced a variety of events during its vast history. When studying such an area, either in the field or on a geology map, ask yourself two broad questions:

1. what events have occurred here?
2. in what order did these events take place?

By fitting the answers together you will interpret the area's geological history.

Several 'laws' have been written to help you sort out the relationships of different rocks and structures. Although they have grand titles, each law is based on simple fact.

Law of superposition. 'If one sedimentary rock lies above another then the upper rock is younger than the lower one.' (An exception to this occurs on the overturned limb of an overfold: see Fig. 10.8.)

Law of cross-cutting relationships. 'A rock or structure must be younger than any rock or structure which it cuts across. For example a dyke must be younger than the sediments pierced by it.

Law of included fragments. 'If a fragment of one rock is found included in another then the included rock must be the older.' For example, a pebble of basalt in a conglomerate shows that the basalt existed before the conglomerate was laid down.

Fig. 11.3 shows a vertical quarry face. You can use the laws mentioned above to work out the geological history.

The age order of the sediments is: A followed by B, followed by C and finally D (law of superposition).

Fig. 11.2 The Brokram Breccia near Appleby, Cumbria. This Permian age rock clearly illustrates the law of included fragments by containing pebbles of Carboniferous limestone complete with identifiable fossils

The dyke is younger than the sediments because:
- it 'cuts' through them (law of cross-cutting relationships);
- xenoliths of sediment are found in the dyke (law of included fragments).

The fault is younger than the sediments and dyke because it 'cuts' through them all (law of cross-cutting relationships).

Hence the geological history shown by Fig. 11.3 is:

(a) Deposition of sediment A   Oldest event
(b) Deposition of sediment B
(c) Deposition of sediment C
(d) Deposition of sediment D
(e) Intrusion of dyke
(f) Formation of fault   Youngest event

Note. Many geologists prefer to write histories with the oldest event at the bottom of the list and the youngest at the top. In this way the true position of the rocks (one above the other) is suggested (as in the geological column).

Fig. 11.3 Relative ages of rocks and structures (see text for details and interpretation)

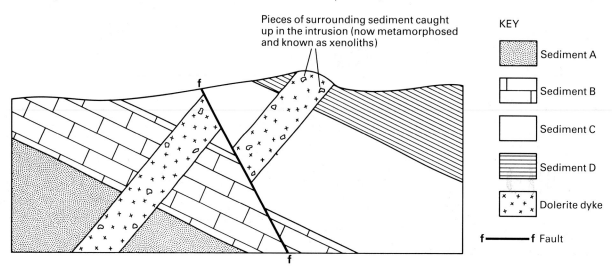

Pieces of surrounding sediment caught up in the intrusion (now metamorphosed and known as xenoliths)

KEY

Sediment A

Sediment B

Sediment C

Sediment D

Dolerite dyke

f———f Fault

Mapwork and geological histories

The most interesting geological histories to work out are those based on a sequence of rocks you have studied in the field. However, they can also help to change a geological map from a mixture of lines and symbols into a meaningful story. Many geological examinations ask you to write a history based on evidence from a map and section. An example of the technique is given below. It is based on Map 6 and Section 6.

Map 6

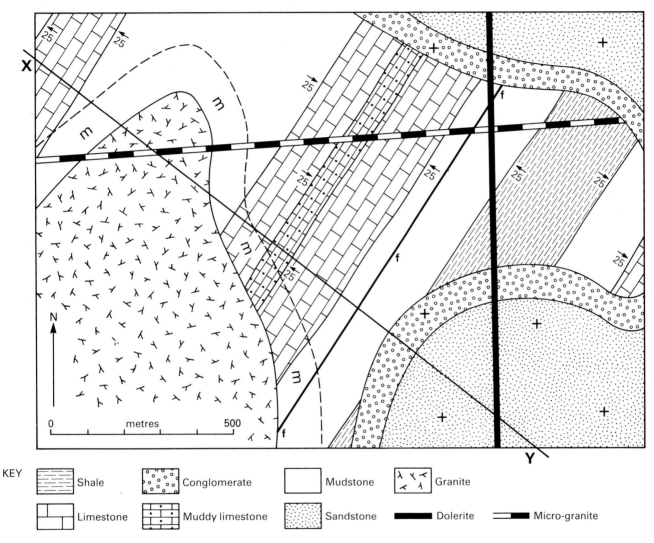

KEY

Shale	Conglomerate	Mudstone	Granite
Limestone	Muddy limestone	Sandstone	Dolerite	Micro-granite

Section 6

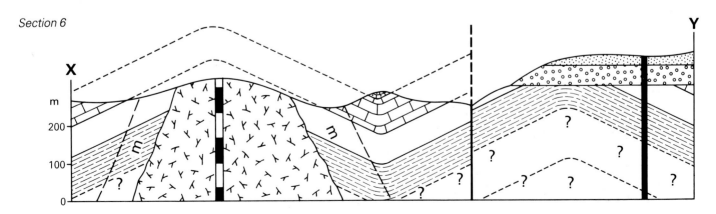

Brief geological history of Map 6
(a) Deposition of shale (oldest rock shown).
(b) Deposition of mudstone.
(c) Deposition of limestone.
(d) Deposition of muddy limestone.
(e) Folding of shale, mudstone and limestones.
(f) Fault affects sequence of folded sedimentary rocks.
(g) Intrusion of granite batholith with metamorphism of surrounding rocks.
(h) Intrusion of micro-granite dyke.
(i) After period of erosion, conglomerate deposited unconformably above other rock types.
(j) Deposition of sandstone above conglomerate.
(k) Intrusion of dolerite dyke (youngest rock shown).

This history is correct but makes rather uninteresting reading. With some care and geological thought you can add detail and really tell the story properly. Compare the full history below with the outline above. Obviously the full history would gain more marks in a geology exam.

Full geological history of Map 6
(a) The oldest rock exposed in the area is shale. No doubt other, older, rocks exist at depth below this.
(b) The shale was overlain by mudstone.
(c) Limestone was deposited above the mudstone.
(d) This limestone was overlain by muddy limestone.
(e) The sequence of sediments listed in parts (a) to (d) show that the area was once covered by sea and that different sediments were formed as the environment changed. For example, the deposition of limestone suggests warmer shallower conditions than those occurring earlier when the shale was laid down.

(f) The shale, mudstone and limestones were then folded into a series of symmetrical anticlines and synclines. The fold axes run NE to SW while the limbs dip at 25° to the NW and SE.
(g) A fault then occurred parallel to the fold axes. The section shows a small downthrow of about 35 m on the NW side but it is possible the fault may also have some horizontal or tear movement.
(h) A large granite batholith then intruded into the folded faulted rocks causing contact metamorphism. The aureole is 100 m wide and should contain hornfels, spotted rock and marble.
(i) An east–west trending dyke of micro-granite then intruded across the region.
(j) The folding, faulting and batholith are probably all related to a period when earth forces were affecting the area (presumably during an orogeny). The region must also have been uplifted at this time because a period of erosion followed.
(k) Erosion wore down the folded sediments and exposed the batholith before a new period of deposition began.
(l) The younger sequence of sediments began when a conglomerate was laid unconformably on the earlier rock types. This suggests that the sea had flooded over the region once more.
(m) Sandstone was laid down above the conglomerate.
(n) The conglomerate and sandstone are still horizontally bedded so it appears the area has not been affected by a second period of folding.
(o) A dolerite dyke was intruded in the east of the area. It has a north–south trend and is the youngest rock shown on the map.
(p) Finally uplift and erosion have caused the rocks to outcrop at the surface of the area.

Questions

1 Study the geological map below and answer the questions which follow. Write the letters (A,B,C,...) of the correct answers in your notebook.

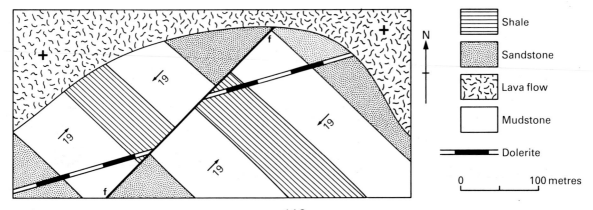

112

(a) What type of fold structure is shown by the sediments (they have not been overturned)?
   A symmetrical syncline   B asymmetrical syncline
   C symmetrical anticline   D asymmetrical anticline
(b) What type of fault is shown?
   A normal   B reverse   C tear   D thrust
(c) What is the displacement of the fault?
   A 10 m   B 100 m   C 50 m   D 150 m
(d) Which is the oldest rock type shown on the map?
   A dolerite   B shale   C mudstone   D sandstone
(e) Which rock lies above an unconformity?
   A sandstone   B lava   C mudstone   D shale
(f) If the sediments are known to be Ordovician age and the fault moved during the Devonian, what age is the dyke likely to be?
   A Carboniferous   B Tertiary
   C Jurassic   D Silurian
(g) If radiometric dating shows that the lava flow is 295 million years old, during which period was it formed?
   A Carboniferous   B Silurian
   C Cretaceous   D Triassic
(h) What is the trend of the fault plane?
   A N.W./S.W.   B N.E./S.W.
   C N.W./S.E.   D N.E./S.E.
(i) What is the dip of the lava flow?
   A 90°   B 45°   C 0°   D 120°

2 If a bed of sedimentary rock dips at 45° to the north, what is the direction of strike?
   A east   B north–south   C west
   D east–west   E south–east

3 Which one of the following could not be caused by compression?
   A anticline   B nappe   C normal fault
   D thrust fault   E reverse fault

4 Study the geological cross-section below and answer the questions which follow.

(a) What name is given to the type of intrusion labelled A? Name the rock type you would expect to find here.
(b) What name is given to the type of intrusion labelled B? Name the rock type you would expect to find here.
(c) Describe the fold types shown at C and D.
(d) Name the feature labelled E–E and explain how it would have formed.
(e) Describe the differences you would expect between specimens of rock collected from points F and G; explain what would have caused these differences.
(f) What evidence suggests that the dark muddy limestone is considerably older than the shelly limestone?
(g) Hard 'baked blocks' of mudstone have been found in the large intrusion at point H; what term is used for such 'blocks' and what does the presence of them tell you about the relative age of the mudstone and this intrusion?
(h) Radiometric dating has shown the age of the coarse-grained igneous rock to be 225 million years and that of the fine-grained igneous rock to be 65 million years; during which era were the conglomerate and shelly limestone deposited?
(i) Write a brief geological history of the rock sequence shown in this section.

5 Draw a rectangle 14 cm wide by 4 cm high. Use this as a framework to draw a labelled cross-section through an area with this geological history:
'The oldest rocks in the area are a sequence of sedimentary beds (limestone followed by sandstone followed by shale). These were folded into a series of symmetrical anticlines and synclines whose limbs dip at 40°. A dolerite dyke then intruded along the fold axis of one of the anticlines. The top of this folded series underwent erosion before horizontal beds of desert sandstone (with large-scale cross-lamination) were deposited unconformably above.'

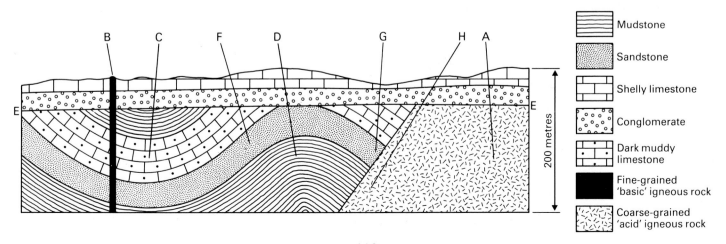

12 Fossils

What are fossils?

Fossils are preserved evidence of prehistoric life. The term 'fossil' does not just refer to the remains of living organisms; it includes any other information such as footprints or feeding trails, which happen to have been recorded in rocks. Articles produced by human development over the last 10 000 years should not be called fossils. Stone axes, remnants of early clothing, etc. are considered to be part of history rather than geology.

Because deposition causes the burial of organisms, nearly all fossils are found in sedimentary rocks. In general, fine-grained sediments are most likely to preserve the best fossils because they can cover, infill and protect small details more closely. Also, since fine sediment tends to accumulate in quiet conditions, there is less chance of current action removing and/or damaging specimens in this type of environment. Some volcanic ashes may bury and fossilise organisms but usually the heat of igneous activity destroys living material.

In normal circumstances, soft body tissue decays quickly or is eaten by scavengers. For this reason, it is usually only the **hard parts** of an organism which are preserved. Such hard parts include the following materials.

1. Shells or external skeletons (**exoskeletons**). Many invertebrates (animals without backbones) have structures of this type which are usually made of **calcium carbonate** ($CaCO_3$). Other materials include the complex organic substance known as **chitin** (used by trilobites: see page 119) and **silica** (SiO_2) used for the skeletons of sponges.

Fig. 12.1 Dinosaur footprint: an example of a trace fossil

Fig. 12.2 Discovering a fossil

2. Internal skeletons; especially vertebrate bones and teeth made of **calcium phosphate**.
3. Woody material from some plants in the form of **lignin**.

Since most fossils show only the hard parts of an organism, geologists are often left with the problem of working out what the missing soft body features could have been like. An even greater problem comes from the fact that many organisms had no hard parts and so have left no fossils to show that they ever existed. The lack of fossils in Pre-Cambrian rocks is probably because many of the earliest forms of life were completely soft bodied.

The formation of fossils

Some fossils are found in recent materials which have not yet been compacted into solid rock. Examples include **bivalve** shells (page 123) from the 1 million year old Crag Deposits in East Anglia and the bones of Quaternary mammals preserved in some peat bogs. In such cases the hard parts have been protected from decay but practically no chemical alteration or hardening has taken place.

In really exceptional circumstances, whole, almost unaltered, organisms from the very recent past have even been preserved. For example, wooly mammoths were found in Siberia still in a deep-frozen state from the effects of the Ice Age.

Materials such as ice, peat or even loosely cemented shell beds are unlikely to survive for long periods of geological time, however. For fossils to remain protected for many millions of years, specimens need to be preserved in solid rock. When this takes place the processes which **lithify** the rock (see page 77) also alter the fossils within it. The type of alteration includes the following.

Carbonisation. This affects organic substances such as **lignin** and **chitin**. During compaction, chemical changes release oxygen, hydrogen and nitrogen from the original fossil material to leave the remaining part rich in carbon. The result is the type of blackened imprint often left by fossil plants (see Fig. 12.48 on page 136). In very rare cases soft body tissues may leave a thin carbon film and preserve details other than the more normal remains of hard parts.

Petrifaction/replacement. This is a result of mineral solutions seeping through a sediment. Sometimes the original fossil material is strengthened by new minerals precipitating to infill small porous spaces. In other cases (see Fig. 12.3) the original material is dissolved away and entirely replaced by the crystallisation of a new mineral. The commonest petrifying and replacing minerals are those which act as cements (page 66) in sediments: for example, **calcite** and **silica** (quartz). In rarer cases fossil material may be replaced by **pyrite, haematite** or **limonite**.

The fossil record

Some organisms are much more likely to be fossilised than others. Follow the flow chart in Fig. 12.5 to see the many things that may prevent a fossil from being formed, preserved and found.

As geologists try to interpret the story of life in the past they can only use evidence that has actually been recorded. It should always be remembered that the **fossil record** presents two major problems.
1. It is a very **incomplete** record. Estimates suggest that of all the different species that have ever lived on earth less than 1 in 5000 has left fossil remains to show it ever existed. Note that this figure is for species not individual organisms!
2. It is a very **biased** record. Most fossil species are of small shelled invertebrates which lived fixed to, or burrowing in, shallow sea floor sediments. All other types of organism (soft-bodied, land-living, etc.) are only preserved in extremely small numbers.

A Geologist's Last Request?

Cremation – It is not for me.
Place my bones in a tranquil sea,
In fifty million years from now,
A fossil Me could be famous – Wow!

Fig. 12.3 How a mould and cast of a fossil can be formed. *Note*: The completely empty mould shown in stage 2 may never actually exist because it is possible for the original material to dissolve at the same time as the new mineral is precipitating

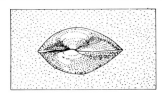

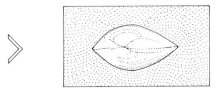

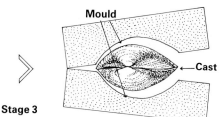

Stage 1 Organism (hard parts) buried in sediment

Stage 2 Original material dissolves to leave an exact mould of itself

Stage 3

New mineral infills mould to produce a cast. Both mould and cast are fossils and each shows detail of original organism

Fig. 12.4 Environments where fossils form

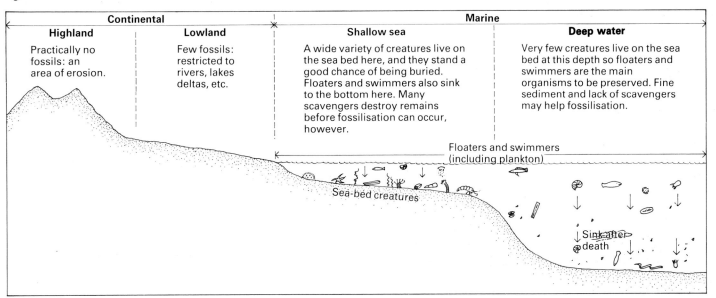

Continental		Marine	
Highland	**Lowland**	**Shallow sea**	**Deep water**
Practically no fossils: an area of erosion.	Few fossils: restricted to rivers, lakes deltas, etc.	A wide variety of creatures live on the sea bed here, and they stand a good chance of being buried. Floaters and swimmers also sink to the bottom here. Many scavengers destroy remains before fossilisation can occur, however.	Very few creatures live on the sea bed at this depth so floaters and swimmers are the main organisms to be preserved. Fine sediment and lack of scavengers may help fossilisation.

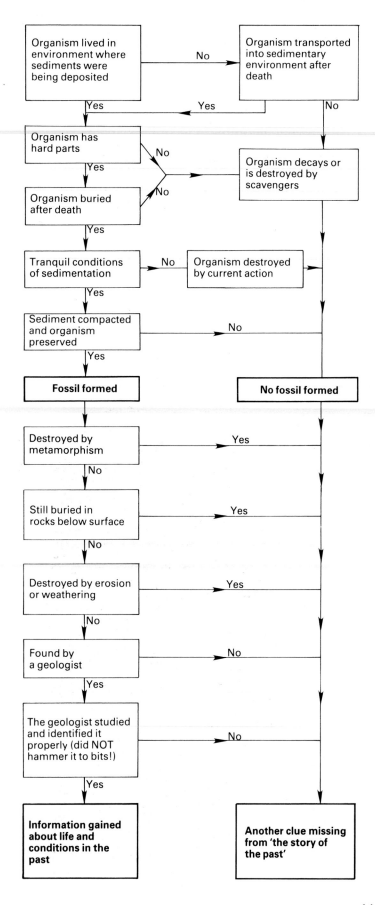

Fig. 12.5 Flow chart of fossilisation

The classification and naming of fossils

All organisms, whether fossilised or living, are divided into groups according to their structures. Classification begins by separating into two **kingdoms** (plants and animals) and then dividing these into **phyla** (singular: a **phylum**). Each phylum contains a range of organisms which have a major feature in common. For example, the phylum chordata contains all the animals with backbones. As the following example shows, further sub-divisions can be made, each using a different feature to define that group.

Kingdom:	Animal
Phylum:	Chordata (backbones)
Class:	Mammalia (mammals)
Order:	Primates (includes monkeys, apes and humans)
Family:	Hominid (human-like)
Genus:	*Homo* (human)
Species:	*Homo sapiens* (modern human)

The smallest unit of classification is a **species**. This could be defined as 'a type of organism that is recognisably different from all other types'. Unless you become an expert, it is usually sufficient to name fossils according to their **genus**. For example, a name such as *Paradoxides* (see page 120) covers a genus of closely related trilobite species.

The study and use of fossils

The study of fossils is called **palaeontology** (from Greek words meaning ancient life). Despite its problems, the fossil record can be used to help geologists:
- understand the theory of **evolution** (how life forms have developed);
- date and correlate rock sequences;
- interpret the environments of the past.

We must look at each of these uses in more detail.

Fossils and evolution

Fossil evidence is vital to scientists trying to explain the way life on earth has developed and changed. The most famous person to write on this subject was **Charles Darwin** who, in 1859, published a theory known as **evolution by natural selection**. His ideas were based on the following points.

During its lifetime an organism is capable of producing a large number of offspring. For example, a single tree may shed thousands of seeds, and one pair of animals may produce many young. Darwin argued that such overproduction had two important effects.

1. The large number of offspring have to compete

with each other for survival. As a result many will die and only the 'fittest' grow to maturity and produce the next generation. Note that competing or being the fittest does not necessarily meaning fighting and killing your opponents. It is a matter of being better equipped to survive in your environment.

2. With so many offspring being produced there is a fair chance that some will have slightly different characteristics from all the others. If these characteristics give a particular organism an advantage over its competitors then that organism is more likely to survive and breed. If the advantage is then passed on to (**inherited** by) the next generation, they too will be more likely to survive and breed. Eventually those organisms with the advantage will outnumber and replace those without it. In this way organisms tend to **evolve** (change with time) **to become better adapted to their environment**.

An example of an **adaptation** is known from studies of peppered moths. Normally these insects have a light body colour and are difficult to see (well camouflaged) against lichen-covered tree bark. (Camouflage protects a creature and helps it to survive.) In 1848 a dark variety of peppered moth was first noticed in Manchester. This was better suited to camouflage against the blackened trees of an industrial city. By 1895, 98% of the peppered moths in this area were of the dark type. This example also shows why the term **natural selection** is used for this process. A **natural** activity (in this case, birds looking for moths to eat) **selects** which organisms are likely to **survive** and which are likely to become **extinct** (die out).

There is also fossil evidence which supports the idea of evolution by natural selection. The series of changes shown by some Mesozoic echinoids (page 129) can be explained in terms of their becoming progressively more suited (**adapted**) to a burrowing environment. The evolution of the horse from a small woodland creature to a large grazing animal is well recorded in Tertiary fossils, and can be logically linked to the evolution of grasses which took place at the same time.

The possibility of change and the development of new characteristics comes from the way reproduction takes place. Sex cells contain a complex chemical 'code' which controls how offspring develop. Since the code carries millions of separate details there is always a chance that something may not be copied exactly and a **mutation** (change) may occur. If the mutation is severe (for example, a fox born with only three legs), it is unlikely to survive. But if the mutation provides an advantage (for example, a fox with a better sense of smell) then it will be supported by natural selection and will provide another small step in the processes of adaptation and evolution.

Since mutation and natural selection only work one step at a time, evolution is a very slow process. This idea should present no problems to geologists who are used to dealing in vast amounts of time. It is thought that the evolution of the horse, for example, took place over 50 million years and involved some 15 million generations. There is a possibility that more rapid 'bursts' of evolution may have taken place at certain times, but this is not fully understood.

It is not certain how far back evolution goes. Since all vertebrates (animals with backbones) have a similar body structure, can we assume they have all evolved from one common **ancestor**? Even more interesting is to consider whether all species can be traced back to just one original 'spark of life'. The question of tracing our human ancestors was the one which landed Darwin in the most trouble of all: it was considered a total insult when he suggested humans were related to monkeys.

Fig. 12.6 A cartoonist's view of Darwin and his theory of evolution (1860s)

Use of fossils in dating and correlation

When you watch an old television programme or film do you ever try to work out when it was made? If you do, you probably decide by looking at the fashions worn or perhaps the types of car shown. Features such as these help us to date scenes from the past because

● new styles keep replacing the old;
● each particular time has its own distinctive set of fashions and vehicles.

In a similar way, fossils can help us to date rocks. Life on earth has constantly changed, with new species evolving as older ones become extinct. As a result, each geological unit of time has a distinctive set of fossils which date it by distinguishing it from all other times.

The most precise dating relies on **zone fossils**. These are particular organisms which existed for a relatively short amount of time (note that short in a geological sense may still mean 10 million years or more!). Fig. 12.7 illustrates the idea of dating by zone fossils. Study the diagram as you consider these possibilities.

1. You find a rock containing fossil a. This does not allow very precise dating. This species had a large time range and the rock could therefore have been formed during any of the five time units.

2. You find a rock containing fossil b. This is a good zone fossil because it has a limited time range. The rock can be dated as having been formed during time unit 4.

3. You find fossil c and fossil d in the same rock. A little thought shows this must have been formed in time unit 3 since that is the only time when these species were alive together (that is, the ranges overlapped). *What age is a rock containing fossils c and e?*

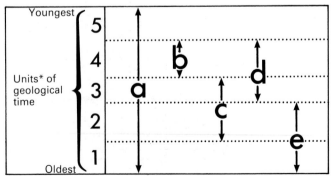

* These time units may be of any length: whole periods (say Cambrian to Carboniferous) or small divisions of one period.

Fig. 12.7 Dating rocks by fossil content. The arrows show the times when each different fossil species a to e was living (that is, its time range)

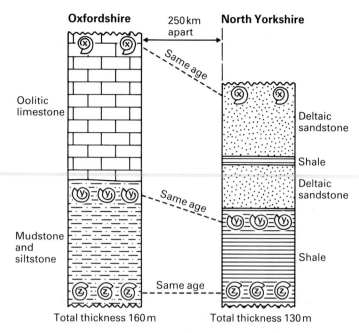

⑨ ⓥ ⓖ Three species of ammonite zone fossils

Fig. 12.8 Example of zoning and correlation (see text for details)

As well as a short time range, good zone fossils also need to be widely found in different rock types from different areas. These factors allow them to be used in **correlation**.

Correlation is the technique of discovering what 'connections' (if any) exist between separate exposures of rock. An example of how this may be achieved is illustrated by the rock sequences in Fig. 12.8. These are from two different parts of Britain and contain differing thicknesses of different sedimentary rocks. At first sight there may seem to be no connection between them. However, the presence of the same zone fossils in each shows that there is, in fact, a connection because the two sequences must be equivalent in age. In other words, although both sets of rocks were produced in different environments of deposition, they were both forming during the same part of the Jurassic period.

Notice that in our example, **ammonites** are used for the zoning and correlation. The most important thing about these creatures (see page 126) was their free-swimming way of life. This meant that they were widely distributed, and after death they sank and became fossilised in many types of marine environment. They can therefore be used to correlate a wide range of sedimentary rocks. Organisms which lived on, or fixed to, the sea floor (**corals**, for example) do not make such useful zone fossils because they could not become so widely distributed.

Before going on to the next section you should re-read pages 88 to 90. Remember that zone fossils only provide a **relative** date, and correlation is also possible using **radiometric** methods.

Fossils and past environments

The section on evolution explained how organisms become adapted to suit their environment and way of life. A biologist, studying living things, can show exactly how organisms are related to environments. But a geologist, working from fossils, has a more difficult task because part of the information will be missing. For example, it may be assumed, from the shape of a fossil shell, that this particular animal lived half buried in sediment. However, other adaptations to this way of life, such as camouflage colours or perhaps eyes raised on stalks, will probably never be known about because they were not fossilised.

The best fossils for indicating past environments are those with close living relatives. In these cases we *know* the present environment and can therefore *infer* the past one. For example, because modern coral reefs only live in clear shallow tropical seas we *assume* fossil reefs must have lived in very similar conditions. Other examples of fossils which are linked to specific environments will be given in the sections on individual fossil groups which follow.

Only fossils preserved in their original environment can give correct information. If a creature lived in one environment but for some reason (for example, current action) was transported into another area before being fossilised, then the information from it would be misleading. In general the best environment indicators are organisms which lived fixed to or burrowed into sediment.

The interpretation of ancient environments gives a picture of changing climates and various distributions of land and sea. This study is sometimes called **palaeogeography** (the geography of the past) and can be illustrated on special maps (for example, a palaeogeographical map of Britain in Devonian times as shown on page 160).

Of course, information about the past does not only come from fossils. Many rock types and structures are also good environmental indicators. For example, red dune bedded sandstones are typical of past desert conditions.

Trilobites

- Trilobites are an extinct group of invertebrates which belong to the same phylum as insects, centipedes and crabs. All these animals have segmented bodies (that is, made up of definite, distinct units) supported by an exoskeleton of chitin.
- The trilobite exoskeleton has three main parts (tri means three): the **cephalon, thorax** and **pygidium**, which correspond to the head, body and tail (see Fig. 12.9).
- During life the exoskeleton was flexible with each segment of the thorax (pleura) able to move against its neighbours. Fossils of curled up trilobites are occasionally found.
- The exoskeleton could not grow with the animal and so periodically the old one was discarded (moulted) and a new one formed. This practice increased the number of exoskeletons available for fossilisation.

Usefulness to a geologist

Trilobites first appeared during the Cambrian and are the most important zone fossils for rocks of this period. They remained fairly common through the Ordovician, Silurian and Devonian but became increasingly rare after that. The last few species of trilobite became extinct during the Permian.

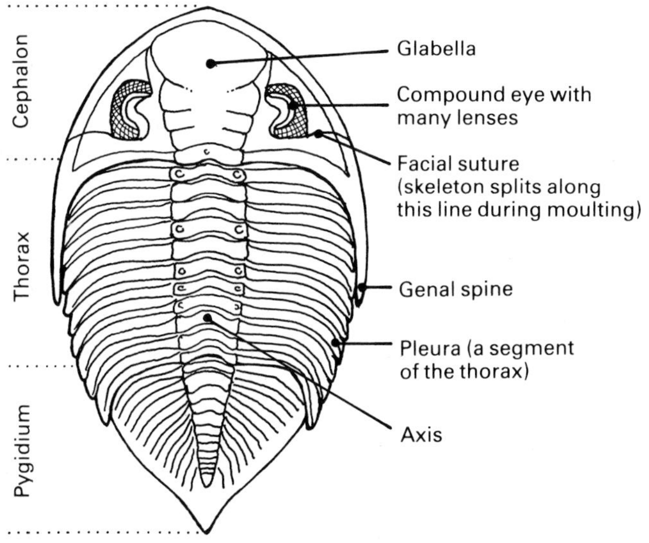

Cephalon

Glabella

Compound eye with many lenses

Facial suture (skeleton splits along this line during moulting)

Genal spine

Thorax

Pleura (a segment of the thorax)

Axis

Pygidium

Fig. 12.9 Features of a typical trilobite exoskeleton. Specimen shown is *Dalmanites* (Silurian): actual size 5 cm

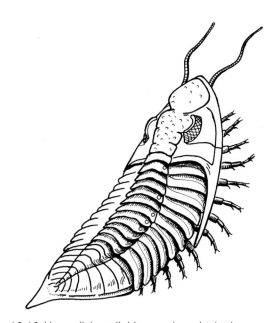

Fig. 12.10 How a living trilobite may have looked

119

Interpreting trilobites

Since there are no living examples to learn from, trilobites present an interesting challenge to geologists. By studying the variation of form shown by fossil exoskeletons can we work out trilobite life styles?

Some known facts help. All trilobites were marine animals and, thanks to the occasional preservation of soft tissues, the general features of the living body have been discovered (see Fig. 12.10). The typical way of life seems to have involved walking and feeding on the surface of sea-floor sediment. The exoskeleton would provide protection from above, while the protruding, upward-facing eyes (shown by most types) gave warning of predators.

Variations on this life style can also be suggested.
- Some species have no eyes: does blindness suggest a life in the deep dark zones of the ocean?
- Some examples have a smooth outline, close fitting pleurae and a scoop-shape cephalon: could these be adaptations for tunnelling?
- Could the reduced spiny exoskeletons of some types have been an adaptation to reduce weight and allow them to swim freely?

Like other geologists, you must draw your own conclusions from the evidence of the specimens you see.

Fig. 12.12 Trilobite fossil. Specimen shown is *Calymene* (Silurian): actual size 6 cm

Fig. 12.11 Trilobites from the Lower Palaeozoic, showing variation in size, shape and possible adaptions

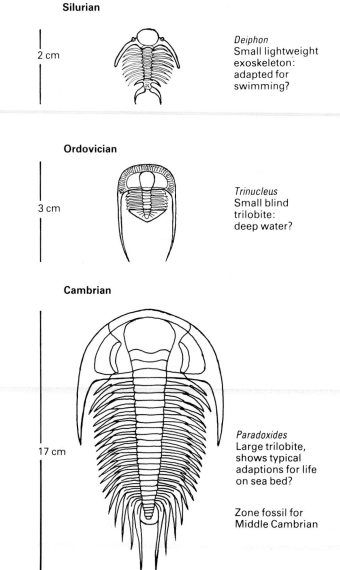

Silurian

2 cm

Deiphon
Small lightweight exoskeleton: adapted for swimming?

Ordovician

3 cm

Trinucleus
Small blind trilobite: deep water?

Cambrian

17 cm

Paradoxides
Large trilobite, shows typical adaptions for life on sea bed?

Zone fossil for Middle Cambrian

Graptolites

- Graptolites are an extinct group of creatures which are most commonly found fossilised in dark shales. At first sight they look similar to small fret-saw blades.
- Each fossil is actually the exoskeleton of a colony of animals whose bodies were housed in tiny cup-like **thecae** arranged in rows along **stipes** (see Fig. 12.13).
- Without living examples to study it is impossible to be certain about the mode of life of a graptolite colony. However, there is evidence to suggest that most species drifted in ocean waters attached to floating objects (seaweed?) by means of a thread-like **nema**.

- Some geologists think graptolites may be distantly related to vertebrates. This is because each animal in a graptolite colony was linked to a common canal running along the length of the stipe (like links to a central 'spinal column'?)

Usefulness to a geologist

Graptolites first appeared during the Cambrian and became extinct in the Lower Carboniferous. They are most commonly found in rocks of Ordovician and Silurian age where, for the following reasons, they make excellent zone fossils.
1. They evolved rapidly, so each particular species only lived for a relatively short time. The Ordovician period, for example, has been subdivided into 14 separate time zones, each of which contains distinctive graptolites.
2. Graptolites were numerous and free-floating. This allowed their fossils to be widely distributed and so the same species can be found and used when correlating rocks from different continents. The only problem with using graptolites for correlation is that they rarely occur in shallow water sediments. Although their skeletons must have sunk into a wide variety of sea floor environments, such small fragile structures would generally have been preserved only in tranquil areas where there were few scavengers (for example, deep-water environments where dark muds and shales were accumulating).

Fig. 12.14 illustrates graptolite zone fossils from the Ordovician and Silurian periods and shows the evolutionary changes which took place. **Dendroid** graptolites such as *Dictyonema* (Fig. 12.15) had many stipes arranged in a branching pattern. These types remained almost unchanged from the Cambrian to their extinction in the Lower Carboniferous and are therefore less useful for specific dating of rocks.

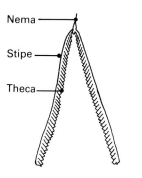

Fig. 12.13 Features of a typical graptolite. Specimen shown is *Didymograptus*: actual size 4 cm

Nema

Stipe

Theca

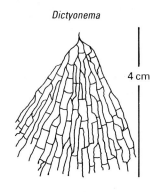

Fig. 12.15 Dendroid graptolite (*Dictyonema*)

Dictyonema

4 cm

Brachiopods

- Brachiopods are an important phylum of shelled invertebrates. Only a few species exist today but the study of these living examples has helped geologists understand more about the numerous fossil types.
- A brachiopod shell is made of two **valves** which are not identical in size and shape. In most species the valves are made of calcium carbonate ($CaCO_3$) and are hinged together allowing them to be opened and closed by a system of muscles (see Fig. 12.16).
- Modern brachiopods can only exist in marine conditions where they lived fixed to the sea bed. It is assumed that most ancient forms lived in the same way.

Usefulness to a geologist

Brachiopods first appear in rocks of Cambrian age. In Britain they are most commonly found in shallow-water sediments deposited during the Silurian, Lower Carboniferous and Jurassic. Despite their abundance during certain periods, most brachiopod species have fairly long time ranges which makes them generally unsuitable for use as zone fossils.

Fig. 12.14 Graptolite zone fossils of Ordovician and Silurian age (average size of specimens 4 cm)

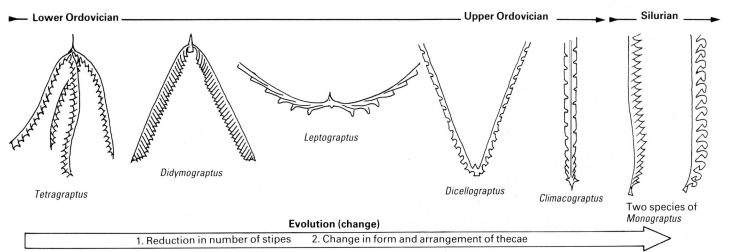

► Lower Ordovician ————————————————— Upper Ordovician ——— ► Silurian ——►

Tetragraptus

Didymograptus

Leptograptus

Dicellograptus

Climacograptus

Two species of *Monograptus*

Evolution (change)

1. Reduction in number of stipes   2. Change in form and arrangement of thecae

121

Fig. 12.16 Features of a typical brachiopod (average size of specimens 2–4 cm)

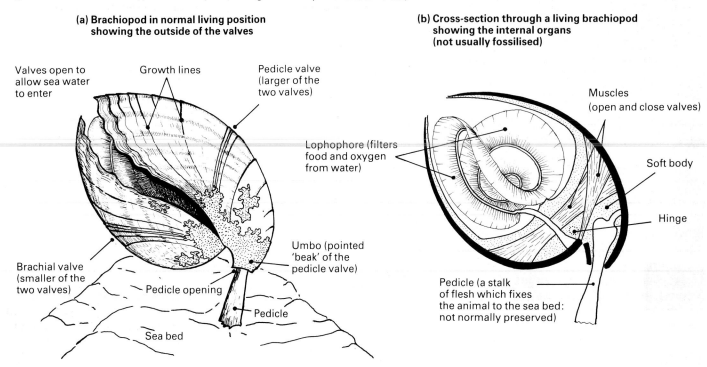

(a) Brachiopod in normal living position showing the outside of the valves

Valves open to allow sea water to enter

Growth lines

Pedicle valve (larger of the two valves)

Brachial valve (smaller of the two valves)

Pedicle opening

Umbo (pointed 'beak' of the pedicle valve)

Pedicle

Sea bed

(b) Cross-section through a living brachiopod showing the internal organs (not usually fossilised)

Muscles (open and close valves)

Lophophore (filters food and oxygen from water)

Soft body

Hinge

Pedicle (a stalk of flesh which fixes the animal to the sea bed: not normally preserved)

Fig. 12.16 shows the features of a typical brachiopod but there were many adaptations to this basic pattern. See if you can tell something about past conditions from the specimens you find. The shapes and features shown by the shell should indicate something about the environment which the brachiopod had evolved to suit. For example, you might conclude that strongly built thick-shelled brachiopods lived in turbulent waters or that small smooth types were adapted to a more tranquil environment. *What environments do you think the brachiopods shown in Fig. 12.17 could have lived in?* Try some geological detective work with specimens you discover on fieldwork trips. Can you, for example, find any relationships between certain brachiopods and certain types of sediment?

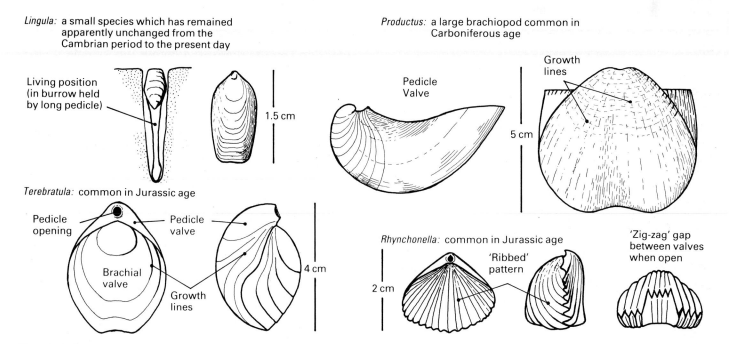

Lingula: a small species which has remained apparently unchanged from the Cambrian period to the present day

Living position (in burrow held by long pedicle)

1.5 cm

Terebratula: common in Jurassic age

Pedicle opening

Pedicle valve

Brachial valve

Growth lines

4 cm

Productus: a large brachiopod common in Carboniferous age

Pedicle Valve

Growth lines

5 cm

Rhynchonella: common in Jurassic age

'Ribbed' pattern

'Zig-zag' gap between valves when open

2 cm

Fig. 12.17 Some examples of fossil brachiopods. Notice several views are needed to show proper shape of each type: follow this idea when drawing specimens yourself

Fig. 12.18 Limestone with brachiopod fossils

Bivalves

● Like brachiopods, these animals have a shell made of two hinged $CaCO_3$ **valves**. However, the two groups are not related since bivalves belong to the **mollusc** phylum. Fig. 12.19 shows how you can tell these fossils apart by noting the shape and symmetry of their valves.

● Fig. 12.20 illustrates a typical bivalve. These shells tend to open up after death so the valves become separated and fossils showing internal features such as **muscle scars** may be found.

● Although the first bivalves appeared during the Ordovician they are generally rather rare in rocks deposited before the Carboniferous. The group increased in number during the Mesozoic and Cainozoic eras and at present there are many living varieties. These include such common creatures as oysters, cockles and mussels.

● Most bivalves are marine animals although a small number of species live in the fresh water of lakes and rivers. A number of types live fixed to a particular place but most are free living and capable of slow movement.

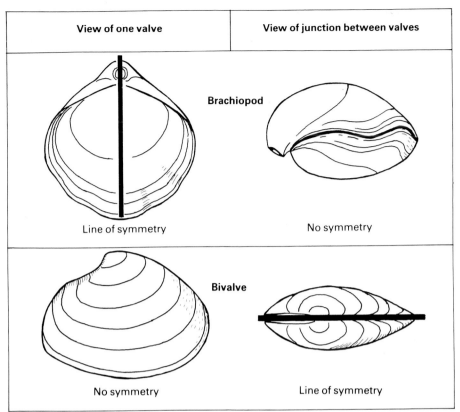

View of one valve	View of junction between valves
Brachiopod	
Line of symmetry	No symmetry
Bivalve	
No symmetry	Line of symmetry

Fig. 12.19 How to tell a bivalve from a brachiopod

Fig. 12.20 Features of a typical bivalve (average size 2–7 cm)

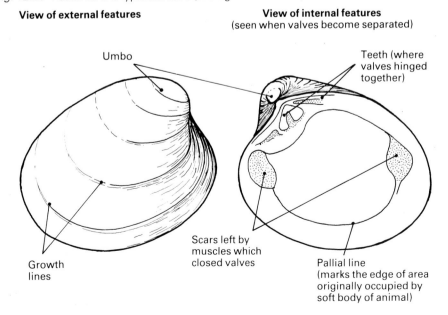

View of external features

View of internal features
(seen when valves become separated)

Umbo

Teeth (where valves hinged together)

Growth lines

Scars left by muscles which closed valves

Pallial line (marks the edge of area originally occupied by soft body of animal)

Usefulness to a geologist

Generally, bivalves are of limited value as zone fossils. The only exceptions to this are non-marine species such as *Carbonicola* which are used to date some Upper Carboniferous sediments (these creatures lived in the lakes and rivers of the coal forest deltas); and a few marine species which date certain parts of the Jurassic period.

The main use of fossil bivalves is to give details of past environments. This is done by comparing their shapes and features with those of living specimens. Since the life style and environment of the living species is known, conclusions can be made about conditions in the past. For example, fossils of *Mytilus* (mussels) would suggest that at the time of deposition a rocky shoreline existed close to that place. See Fig. 12.21 for details of other bivalves and their adaptations to particular environments.

Fig. 12.22 **Pecten** (bivalve) fossils preserved in rock of Jurassic age

Fig. 12.21 Bivalves: ways of life

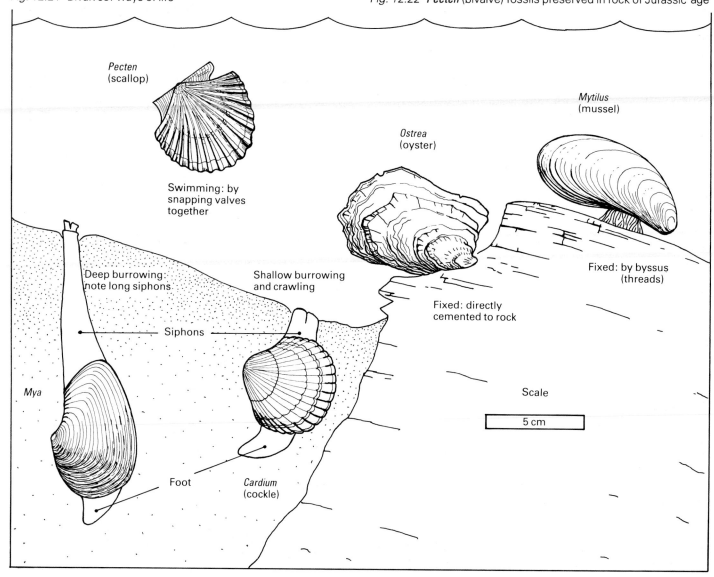

124

Cephalopods

- Cephalopods are also members of the **mollusc** phylum but these animals are much more mobile than bivalves. All modern cephalopods are capable of well co-ordinated movement and live as free-swimming marine animals.
- Living varieties, such as the squid and octopus, show the main features of the cephalopod group. They have a well-developed head, brain, and nervous system and swim either by moving their tentacles or shooting out a jet of water which propels them backwards rapidly. However, only one modern form (*Nautilus*) has the kind of chambered $CaCO_3$ shell which is typical of fossil cephalopods.
- Fossil cephalopods are divided into three sub-groups: **nautiloids, ammonoids** and **belemnites**.

Nautiloids

Studies of the living example *Nautilus* (Figs. 12.23 and 12.24) have helped geologists to understand fossil forms. As the animal grows it adds new larger **chambers** to house its increased body size and seals off the rear of the shell by adding **septa**. In this way a series of inner chambers is formed which are filled with gas to keep the animal buoyant. A thin column of tissue known as the **siphuncle** links the main body to the inner chambers.

Nautiloids first appeared during the Upper Cambrian but most of these early forms (i.e. those found in rocks from the Lower Palaeozoic era) had a straight cone-shaped shell. *Orthoceras* (see Fig. 12.25) shows this shape. It has been suggested that this straight cone structure may have been difficult to balance because, while most of the weight was in the 'head' end, most of the buoyancy was in the chambers towards the rear. Perhaps the later evolution of coiled nautiloids was an adaptation which solved this problem?

Ammonoids

This group evolved (presumably from the nautiloids) during the Devonian and eventually became extinct at the end of the Cretaceous. They are by far the commonest type of fossil cephalopod with many different species preserved in rocks from the Upper Palaeozoic and Mesozoic eras.

Although ammonoids look basically similar to coiled nautiloids, and are presumed to have had a similar way of life, the body and shell structure of these two groups differs. The most obvious difference is seen in ammonoid specimens where the outer part of the shell has not been fully preserved. In such cases lines called **septal sutures** are visible which show where the septa were joined to the inner surface of the shell. While nautiloids always have a set of simple

Fig. 12.23 *Nautilus*

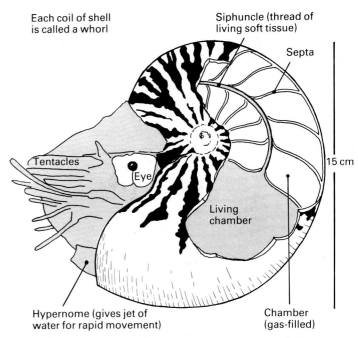

Fig. 12.24 Features of the living cephalopod *Nautilus* (shell sectioned to show internal structure)

Fig. 12.25 *Orthoceras*: a Lower Palaeozoic straight nautiloid

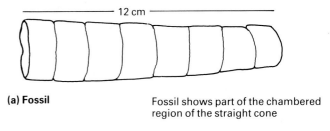

(a) Fossil

Fossil shows part of the chambered region of the straight cone

(b) Reconstruction of living animal

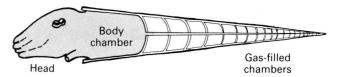

125

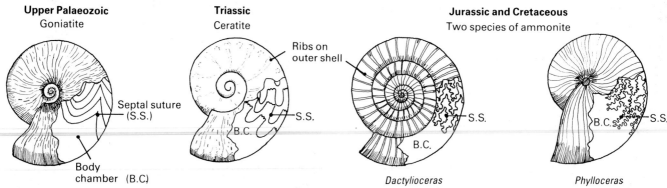

Fig. 12.26 Ammonoid evolution. In each case, part of the outer shell is removed to show the characteristic septal sutures (s.s.) Average size of specimens 6 cm

Upper Palaeozoic
Goniatite

Septal suture (S.S.)

Body chamber (B.C)

Triassic
Ceratite

Ribs on outer shell

S.S.

B.C.

Jurassic and Cretaceous
Two species of ammonite

S.S.

B.C.

Dactylioceras

B.C.

S.S.

Phylloceras

curved septal sutures, ammonoids show more elaborate patterns. It is on the basis of these patterns that ammonoids have been divided into three distinct types (see Fig. 12.26).

1. Goniatites. These have 'zig-zag' suture patterns are only found in Upper Palaeozoic rocks.

2. Ceratites. These have partly 'frilled' suture patterns and occur only in rocks of Triassic age.

3. Ammonites. These have a very complicated pattern of 'frills' on all parts of the suture. They lived only during the Jurassic and Cretaceous but were so common that several thousand species are known. When identifying ammonites you will need to note distinguishing features such as shell ornament (shape and pattern of ribs, etc.) and the style of coiling (is each whorl almost hidden by the next or do successive whorls just touch?).

Belemnites

These cephalopods did not have a totally external shell like the other fossil forms. As Fig. 12.28 shows, they had an internal 'bullet-shaped' **guard** which did, however, possess a typically chambered structure for part of its length. The outer soft body of the creature has occasionally been found preserved as a carbonised imprint, so geologists have been able to reconstruct its likely appearance. Belemnite guards are quite common in some Jurassic sediments but, like the ammonoids, the belemnites died out at the end of the Cretaceous.

Fig. 12.27 Selection of cephalopod fossils (all are ammonites except those labelled B which are belemnites)

Usefulness to a geologist

Ammonoids make particularly useful zone fossils. Like graptolites (page 120) they had
- a relatively rapid rate of evolution which meant that a different and distinctive set of ammonoids lived during each particular time;
- a free-moving way of life which meant that they became widely distributed. In fact ammonoids had an advantage over graptolites in this respect. Because they were larger and more strongly built, ammonoids became preserved in a wider range of rock types (graptolites are rarely preserved in shallow-water sediments).

Marine rocks of Upper Carboniferous age are zoned according to the type of **goniatites** they contain. The Jurassic and Cretaceous periods are sub-divided into numerous smaller time zones on the basis of which **ammonites** occur. Only the Triassic period presents problems for British geologists. Since no true marine rocks were deposited here at that time, **ceratites** are not found in this country.

Belemnites and nautiloids are not generally used for zoning purposes. It is also difficult to use any type of fossil cephalopod for interpreting sedimentary environments. Since they were free swimming they did not actually live where the sediment was being deposited but arrived there (by chance) after death.

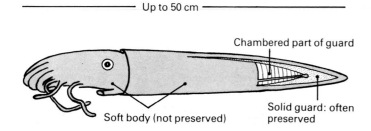

Up to 50 cm

Chambered part of guard

Soft body (not preserved)

Solid guard: often preserved

Fig. 12.28 How a living belemnite may have looked

Gastropods

- Gastropods, like bivalves and cephalopods, also belong to the mollusc phylum. They are the animals we commonly call **snails** and generally have a $CaCO_3$ shell which is composed of just one valve coiled into a tapering spiral (see Fig. 12.29).
- Gastropods are free moving and crawl slowly on a muscular 'foot'. During movement the head, equipped with sense organs, also protrudes from the aperture of the shell.
- Most gastropods live in water where there are both marine and fresh-water species. Some types exist on land.
- Gastropods first appeared in the Cambrian but are rather uncommon in rocks from the Palaeozoic eras. Many more species evolved during the Mesozoic era and, by the Tertiary, gastropods had become quite numerous.

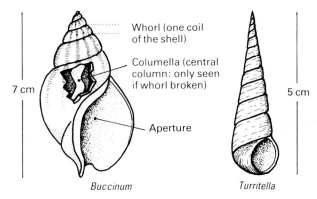

Fig. 12.29 Examples of fossil gastropods

Usefulness to a geologist

Unfortunately, the shape of gastropod shells is not usually a reliable guide to the environments in which they lived. A few types yield information: *Viviparus*, for example (Fig. 12.30) is a genus which still lives today. Since it is known to inhabit fresh water only, fossils of it indicate deposition in a lake or river.

With the exception of Tertiary age sediments, gastropods are not common enough or sufficiently widely distributed to be used as zone fossils.

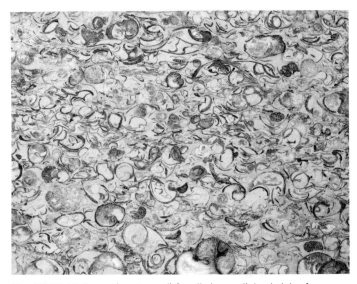

Fig. 12.30 *Viviparus* (gastropod) fossils in a polished slab of Cretaceous fresh-water limestone

Crinoids and echinoids

Crinoids and echinoids both belong to the **echinoderm** phylum which also includes animals such as starfish. All members of this phylum have their body parts arranged in a five-rayed pattern: for example, the five 'arms' of a starfish. Echinoderms also have a shell known as a **test** which is made of numerous interlocking calcite plates. This test is not completely external but, during life, is covered by a thin membrane.

Crinoids

The most commonly fossilised types of crinoid originally lived fixed to the sea bed. They must have appeared almost plant like and are sometimes refered to as **sea lilies**. Fig. 12.31 illustrates a complete crinoid test: notice that there are three main parts.

1. A flexible **stem** made of circular calcite plates called **ossicles**. This may have had an anchoring 'root' but it did not draw up nutrients as a plant would.

2. A hollow **calyx** where the main soft body lived. This was made of calcite plates arranged in multiples of five.

3. A system of five main **arms** which in many species branched out into smaller structures. These parts of the body had grooves running along them and acted as a kind of funnel directing food such as plankton towards the animal's mouth at the top centre of the calyx.

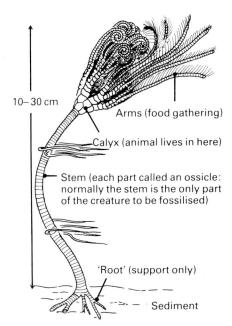

Fig. 12.31 How a living crinoid may have looked

Usefulness to a geologist

Fossil crinoids are mainly found in rocks from the Palaeozoic eras. They first appeared during the Ordovician but are best known from the **crinoidal limestones** of Silurian and Lower Carboniferous age. These rocks are made almost entirely of fragments of crinoid stems.

Crinoids became less common in the Mesozoic, although some rare stemless varieties are used as zone fossils for part of this era. The adaptation towards stemless forms seems to have continued, because most modern species are of that type.

Echinoids

These free-moving marine animals are commonly called **sea urchins**. Fig. 12.33 shows a typical echinoid **test** (shell): notice the following features.

1. A radiating pattern of five **ambulacra** alternating with five **interambulacra** (the singular of these terms is ambulacrum and interambulacrum).

2. Each **ambulacrum** is made of a double row of calcite plates with tiny pores. During life, sucker-like **tube-feet** protrude through these pores to assist movement and respiration.

3. Each **interambulacrum** has a double row of larger calcite plates supporting moveable **spines**. Although spines are rarely preserved on fossil echinoids, you can usually see the **bases** to which they were fixed.

4. The group is divided into two types: regular echinoids and irregular echinoids.

Regular echinoids (Figs. 12.34, 12.35a)

These are circular and have radial symmetry. The mouth is at the centre of the lower surface and the anus at the centre of the upper surface. A strong jaw system allows them to scavenge and graze on plant material as they use their spines and tube feet to move over rocky sea bed areas. Regular echinoids first appeared in the Ordovician and modern species still live today.

Fig. 12.32 Crinoidal limestone

Fig. 12.34 Living echinoid (sea urchin)

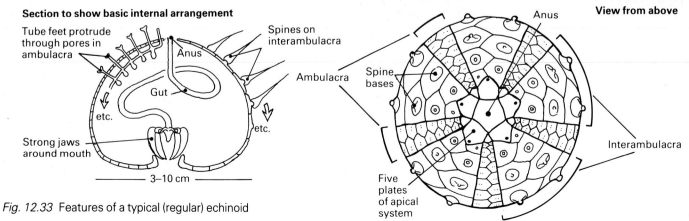

Section to show basic internal arrangement

Tube feet protrude through pores in ambulacra

Anus

Spines on interambulacra

Gut

Ambulacra

etc.

etc.

Strong jaws around mouth

3–10 cm

View from above

Anus

Spine bases

Interambulacra

Five plates of apical system

Fig. 12.33 Features of a typical (regular) echinoid

Irregular echinoids

(Fig. 12.35b)

As you would expect from the name, these echinoids are less symmetrical than regular forms. They generally have either a 'flattened' or heart-shaped test and are assumed to have evolved from regular ancestors before their first appearance in the Jurassic period. The loss of a truly regular shape is thought to be a result of adaptations to different modes of life. This theory is supported by evidence from the irregular echinoid *Micraster*. A series of specimens from Cretaceous sediments shows how this creature became increasingly better suited to a burrowing way of life. As time passed, different species of *Micraster* showed the following adaptations:

- the mouth gradually lost its powerful jaw structure and became located nearer to the front of the test;
- the anus became located closer to the rear of the test.

The total effect of such changes was to make the creature more suited to moving forward within the sediment, but its structure became ever more irregular in the process of this evolution.

Usefulness to a geologist

Echinoids are rare in rocks from the Palaeozoic eras but become increasingly common in those from the Mesozoic era. Both regular and irregular species are used as zone fossils in the Chalk deposits of Cretaceous age.

Fig. 12.35 Examples of echinoids

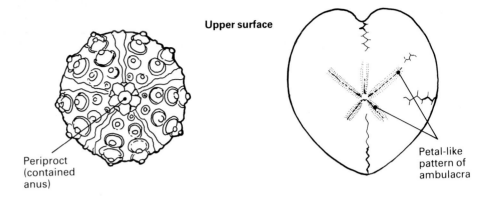

(a) Regular echinoid
Example: *Hemicidaris* (Jurassic)

Upper surface

(b) Irregular echinoid
Example: *Micraster* (Cretaceous)

Petal-like pattern of ambulacra

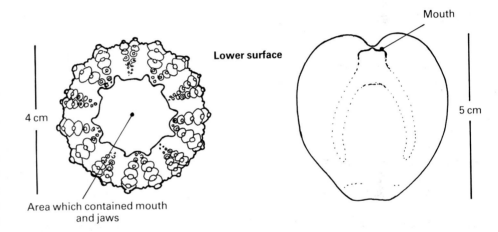

Periproct (contained anus)

Lower surface

4 cm

Area which contained mouth and jaws

Mouth

5 cm

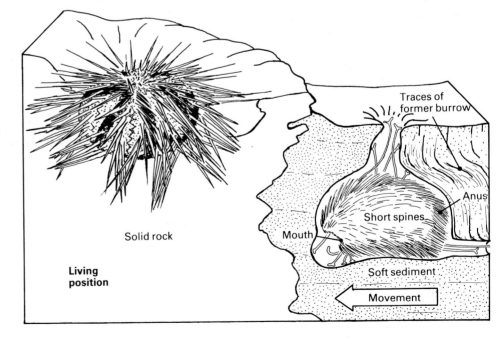

Solid rock

Living position

Traces of former burrow

Anus

Short spines

Mouth

Soft sediment

Movement

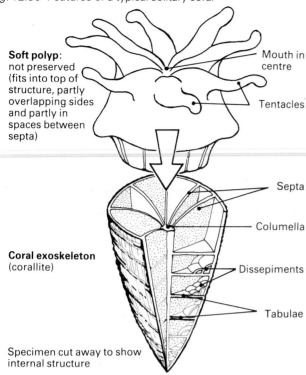

Fig. 12.36 Features of a typical solitary coral

Soft polyp:
not preserved
(fits into top of
structure, partly
overlapping sides
and partly in
spaces between
septa)

Mouth in
centre

Tentacles

Coral exoskeleton
(corallite)

Septa

Columella

Dissepiments

Tabulae

Specimen cut away to show
internal structure

Fig. 12.37 Coral reef
Colonial reef corals need
- **warmth**: only found in tropical seas (30°N to 30°S of equator)
- **light**: will not grow in water more than 20 m deep
- **oxygen**: only survive in water kept turbulent by wave action
- **clear marine conditions**: never develop near river mouths

Corals

- Corals belong to the **cnidaria** phylum which also includes sea anemones and jelly fish. All these animals have a fairly simple body structure consisting of an internal sac supplied with food by a tentacle-surrounded mouth.
- The soft body of a coral is called a **polyp**. This secretes $CaCO_3$ to form a tubular exoskeleton (or **corallite**) which protects the animal and firmly anchors it to the sea floor. As shown by Fig. 12.36, corallites are strengthened and divided by internal walls such as **septa** and **tabulae**.
- Some coral species are **colonial** and have many polyps living in a system of branching or parallel corallites. In the type of environment shown in Fig. 12.37 such colonies may flourish and build into a large strong mass of coral known as a **reef**. **Solitary** corals live as individual animals and can normally survive in deeper or colder environments than the colonial reef builders.

Examples of fossil corals (refer to Fig. 12.38)

The simplest forms belong to a group known as **tabulate** corals. These first appeared during the Ordovician but, in Britain, the best-known examples, such as *Favosites*, come from limestones of Silurian age. Tabulate corals never have septa and are always colonial.

Rugose corals were most important in the Upper Palaeozoic era and are particularly common in Carboniferous limestone. They always contain septa but may be either solitary (example: *Dibunophyllum*) or colonial (example: *Lithostrotion*). Note the rough ridges on the outside of these specimens; this is a typical feature of rugose corals.

Tabulate corals became extinct by the end of the Palaeozoic era and rugose forms became extinct early in the Mesozoic era (Lower Triassic). They were replaced by **scleractinian** corals which are still the main variety today. Although Fig. 12.38 only shows the colonial example of *Thecosmilia*, solitary forms also exist.

Usefulness to a geologist

A mass of colonial corals (found as a **reef limestone**) gives information about a past environment. Since modern reefs need special conditions it is logical to assume very similar conditions must have occurred wherever reefs grew in the past. Additional information about water depth and current direction may come from other creatures such as brachiopods and goniatites which lived, and were fossilised, in the reef area. Some solitary corals are useful zone fossils in time periods where limestone is an important rock: for example, *Dibunophyllum* in the Lower Carboniferous.

Fig. 12.38 Examples of fossil corals

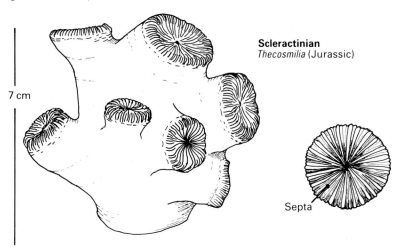

Scleractinian
Thecosmilia (Jurassic)

7 cm

Septa

Rugose
(a) *Lithostrotion* (Carboniferous)

(b) *Dibunophyllum* (Carboniferous)

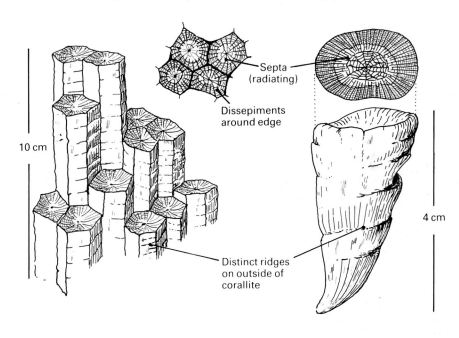

10 cm

Septa (radiating)

Dissepiments around edge

Distinct ridges on outside of corallite

4 cm

Tabulate   *Favosites* (Silurian)

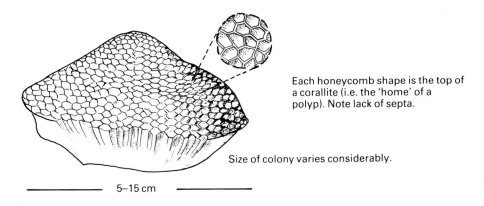

Each honeycomb shape is the top of a corallite (i.e. the 'home' of a polyp). Note lack of septa.

Size of colony varies considerably.

5–15 cm

Fossil vertebrates

Although vertebrates make up only a tiny proportion of fossil specimens it is an interesting task to trace their evolution through the stages of

fish → amphibians → reptiles → mammals and birds.

Fish

The earliest known fish appeared during the Mid-Ordovician. It is not really known what creatures they evolved from but possibly their ancestors were similar to the **lancelet**. This small marine animal exists today and, although it lacks a proper head, fins or limbs, it does possess a simple backbone-like chord. Several fossils of creatures similar to the lancelet have been found in Upper Cambrian sediments.

Early fish had simple sucking mouths, no jaws and only small primitive fins. Both marine and fresh-water species existed and they probably lived close to the bottom, grubbing through sediment for food. For some reason (protection ?) these early fish had heads armoured with thick bony plates.

By the end of the Silurian, jawless fish began to be replaced by more advanced forms called **placoderms**. These had teeth and jaws and were better swimmers despite the fact that most types still carried head armour. Although most placoderms were fairly small, large predators also existed: for example *Dinichthys* (Fig. 12.39).

During the Devonian, fish evolved body structures which are similar to those shown by modern fish. These were of two types, both of which allowed more efficient swimming.

1. Some forms evolved a lightweight skeleton using **cartilage** instead of bone. Sharks and rays are of this type. Unfortunately, cartilage is rarely fossilised so only the teeth of such fish tend to be preserved (see Fig. 12.40).

2. Other forms kept a bone skeleton but developed an internal buoyancy sac called a swim bladder. These types are sometimes known as **bony fish**. Most modern species have this body structure.

Bony and cartilaginous fish both also evolved scales and an efficient fin system. The placoderms could not compete with such developments and so died out during the Carboniferous.

Amphibians

Amphibians (such as frogs) are able to live in water and on land. Movement from water on to land was a very important stage in vertebrate evolution which required two major adaptations: a set of walking limbs and a lung system which could breathe air.

As Fig. 12.41 shows, bony fish have two types of fin structure. **Lobe fins** are rare today but were quite

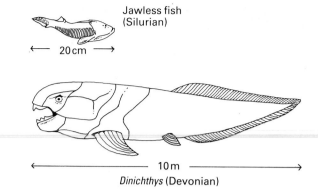

Jawless fish (Silurian)

← 20 cm →

← 10 m →

Dinichthys (Devonian)

Fig. 12.39 Primitive fish

Fig. 12.40 Fossil shark's teeth (size of specimens 2–4 cm)

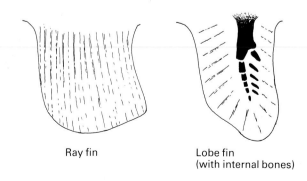

Ray fin

Lobe fin (with internal bones)

Fig. 12.41 Ray and lobe fish fins

132

common in Devonian species. It is quite easy to see how these could have evolved into jointed limbs. The modern lung-fish has lobe fins and is also capable of increasing its oxygen supply by gulping air. Presumably a similar adaptation could have led to the original evolution of lungs and the development of amphibians in late Devonian times. This timing corresponds with the evolution of the first land plants (see page 136). Such plants must have begun to increase the amount of oxygen in the atmosphere and so assisted in the development of animal life on land.

Amphibian fossils are best known from the coalbearing rocks of the Upper Carboniferous. It would seem that the warm swampy conditions of this time were ideal for such creatures, and living in such an environment increased their chances of becoming fossilised. As Fig. 12.42 shows, Carboniferous amphibians were sometimes rather larger than modern types such as toads and frogs.

Reptiles

Unlike amphibians, reptiles have a scaly skin which does not have to be kept constantly moist, and they lay eggs with a protective shell. These adaptations mean that they do not have to live close to water, and reptiles are able to inhabit a far wider range of environments than amphibians can.

The first reptiles appeared towards the end of the Carboniferous period. They increased in number during the Permian and became so numerous in the Mesozoic era (Triassic to Cretaceous periods) that this time is often refered to as the Age of Reptiles.

The best-known Mesozoic reptiles are the **dinosaurs** which include such creatures as the following (see Fig. 12.42).

- *Triceratops* and *Stegosaurus*: armoured dinosaurs which, despite their fierce appearance, were actually plant eaters.
- *Brachiosaurus* and *Diplodocus*: the largest land animals, up to 60 tonnes and 27 m long; they existed on plant material.
- *Tyrannosaurus*: a 7 m high predator with an impressive set of teeth.

Fig. 12.42 Amphibians and reptiles.
(a), (b) are amphibians, (c) – (i) are Mesozoic reptiles

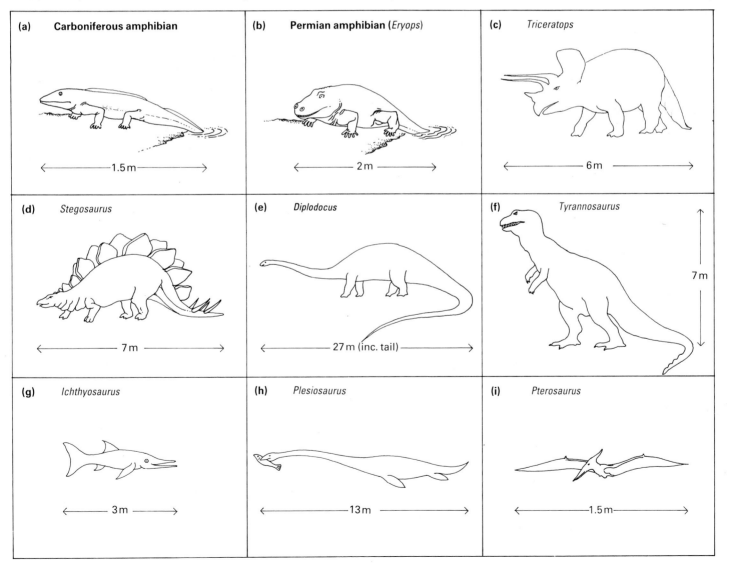

Other reptiles such as *Plesiosaurs* and *Ichthyosaurs* adapted to a free-swimming marine life, while winged varieties such as *Pterosaurs* took to the air. Not all dinosaurs were giants (some were no bigger than chickens) and, despite what is often said about them, they were not stupid. After all, they 'dominated' the land for 150 million years. It is also quite possible that not all Mesozoic reptiles were cold blooded. During the Triassic, one group evolved to produce the earliest mammals.

Although a few species such as crocodiles have survived to the present day, most Mesozoic reptiles became extinct at the end of the Cretaceous period. There are many theories about why this happened. One recent suggestion is that a giant meteorite may have hit the earth at that time. This could have produced enough dust in the atmosphere to change the climate drastically. (You could find out other theories of dinosaur extinction from library books.)

Birds

Birds are possibly the dinosaurs' closest living relatives. Certainly many dinosaurs had a similar hip structure to bird skeletons. Fossil evidence of *Archaeopteryx* (see Fig. 12.43) shows that this Jurassic creature had feathers plus teeth and a bony tail. In other words it seems to indicate a direct evolutionary link by being half bird and half flying reptile. After *Archaeopteryx* (the name means early bird) other species are known to have developed, but fossils are unfortunately very rare.

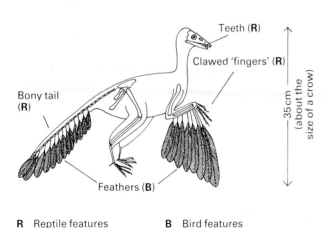

R Reptile features   B Bird features

Fig. 12.43 *Archaeopteryx*

Mammals

Mammals seem to have obvious advantages over reptiles. They are definitely warm blooded and can control their body temperature to survive in environments from the equator to the poles. Their offspring also develop protected within the mother (live birth). It seems strange that after their origins in the Triassic, mammals remained small, rather insignificant, animals until the Tertiary some 130 million years later. The reason, presumably, is that until the majority of reptiles had become extinct, mammals did not have an opportunity to evolve and replace them.

Rapid development of mammals during the Tertiary has led to this time being known as the Age of Mammals, and their present stage of evolution can be clearly seen today. Living forms include marine types (example: whales), flying varieties (bats), plant eaters (deer, etc.) flesh-eating hunters (cats, etc.) and, of course, apes and humans.

Fig. 12.44 Woolly mammoth

Hominids (human-like animals)

The earliest evidence comes from about 60 million years ago when small squirrel-like mammals existed. These were well adapted to life in the forests and, from these beginnings, a number of monkey-like forms evolved between 50 and 25 million years ago. These developments eventually led to the appearance of *Dryopithecus*. Fig. 12.45 shows the evolution from this point onwards. Study it and note the following trends.

- A change to an upright, bi-pedal stance (standing and walking on two legs).
- Increasing brain size (measured from the cavity inside fossil skulls).
- Increasing body weight and size.
- Modifications to tooth and jaw structure to allow for a change in diet as life moved from forests to open ground, and meat eating increased.

Since Darwin's first theories there has always been argument about human evolution. At present there are disputes about the links between *Ramapithecus* and *Australopithecus* (note the time gap between them). Some scientists also think there may not be just one direct line of evolution (that is, at some stages several different hominid types may have all been alive at the same time).

Although *Homo sapiens* is described as the most modern human species, there are a range of forms. For example, *Neanderthal man* (approx 50 000 years ago) had slightly different characteristics to their descendants today.

Fig. 12.45 Hominid evolution

	Dryopithecus	*Ramapithecus*	*Australopithecus*	*Homo erectus*	*Homo sapiens*
Basic Description	Earliest ape	Ape with some pre-hominid features	Human-like ape	Early human	Modern human
Time range	25 to 10 million years ago	15 to 8 million years ago	4 to 1.5 million years ago	1.5 million to 300 000 years ago	Since 300 000 years ago
Brain capacity	Less than 300 cm^3?	?	400–600 cm^3	800–1100 cm^3	1350 cm^3 (average)
Body weight	20–30 kg	?	40–50 kg	50–60 kg	60–70 kg
Stance	Moved on all fours	?	Upright: bipedal (two feet)	Upright: bipedal	Upright: bipedal
Diet (based on tooth structure)	Soft fruit; leaves	Vegetarian: fruit, nuts, seeds	Vegetarian: strong chewing teeth	Vegetable and meat diet (omnivorous)	Omnivorous
Environment	Forest	Open woodland	Open ground near woods	Open ground	Variable!
Technology	No tools	No tools	No real evidence of tools	Stone tools and fire	Ever more complex!
Skull					

Fig. 12.46 Time range of main fossil groups. Dashed lines show full range. Solid lines show
the particular times when each group was fairly common and/or provide useful geological information

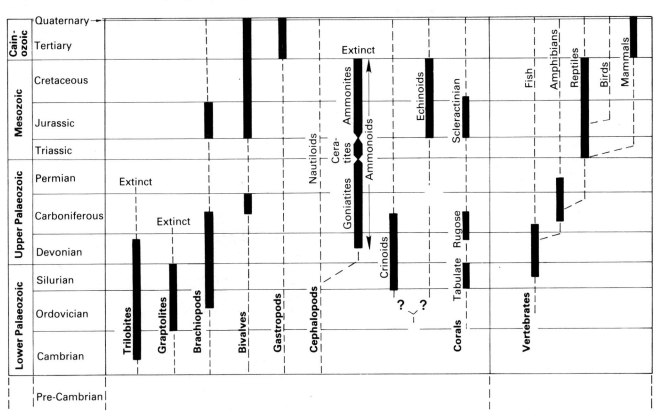

135

Fossil plants

It is thought that the simple types of plant known as **algae** first evolved in the oceans some 3000 million years ago. These were followed by larger, more advanced marine types such as seaweed. But it was not until the late Silurian (400 million years ago) that plants began to grow on land.

The evolution of land plants took as long as this because many adaptations were needed. In the sea, a plant absorbs light, carbon dioxide (CO_2) and nutrients directly from the surrounding water which also gives it support. On land, specialised features are needed to perform each of these tasks. For example, a root is needed to draw up nutrients and water, a stem for support and leaves to absorb sunlight. A land plant also needs a **vascular system** to link all its different parts together and allow materials, in the form of sap, to flow between them.

The first vascular plants to exist on land (in the late Silurian) were primitive ground-creeping types, but other more advanced varieties evolved rapidly during the Devonian. Land plants flourished particularly in the swampy tropical environment of many regions during the Upper Carboniferous, so that this time became the age of **coal forests** (Figs. 12.47 and 12.48).

The majority of Carboniferous plants reproduced by shedding spores (single reproductive cells) but these types mostly died out during the Mesozoic era when seed-producing types evolved. The next stage in development was the appearance of flowering plants which became important during the Tertiary and remain the major group today.

Usefulness to a geologist

Some simple marine plants have helped in the formation of sedimentary rocks such as **chalk** (page 70). **Coal** is the result of the accumulation of material from land plants. Apart from the special conditions of the coal-forming environment (described on pages 71 and 162), fossilised land plants tend to be rare. If you do find one, an obvious conclusion is that the rock containing it must have formed either on land (lake, swamp, etc.) or in a sea area very close to the shore.

Fossil plants are also difficult to classify. For example, if you find pieces of root, stem and leaves how do you know if they were all from one species of plant or if they came from several different ones?

Fig. 12.47 Carboniferous coal forest plants

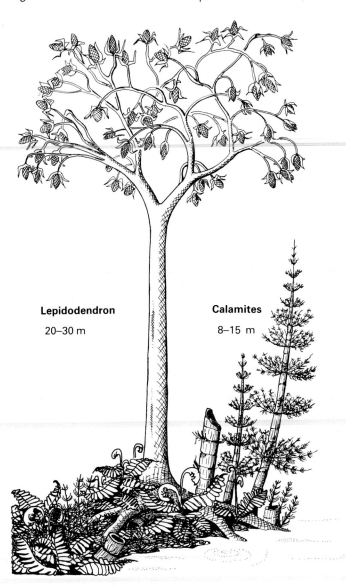

Lepidodendron
20–30 m

Calamites
8–15 m

Fig. 12.48 Selection of plant fossils

10 cm

Fossils: practical work

You cannot fully learn about fossils from diagrams and photographs in books. You really need to see their features and structures in three dimensions by studying real specimens. The best fossil collections can be viewed in museums but you will probably also have a selection of specimens available at school. When studying a fossil you should aim to do the following.

- Identify it: although the exact name is not always needed, you must be able to say to which group it belongs.
- Recognise and name its parts and features.
- Interpret something from it: for example, can you tell which geological age or environment it comes from or what its mode of life may have been?

The ideal way of recording your study is with a scaled labelled diagram. This is not as difficult as it sounds and even poor artists can manage if they go about it in the proper way. As Fig. 12.49 shows, the basic method is to measure the specimen and draw a framework of its dimensions. You can adjust the size of the framework by doubling, halving, etc. all the actual measurements but you must make sure that everything is kept in proportion. Next, sketch in the general outline and take care that, if necessary, this shows the symmetry of the specimen. Only then should you add details and label important features. Finally, do not forget to include a scale arrow showing the fossil's actual size.

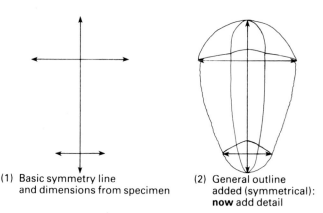

(1) Basic symmetry line and dimensions from specimen

(2) General outline added (symmetrical): **now** add detail

Fig. 12.49 Drawing a trilobite

The test of a good fossil drawing is to ask 'Does it show all the relevant features in their correct sizes and proportions?' This is far more important than whether it is an artistic masterpiece.

When you find fossils during fieldwork, consider if it is really necessary to collect them. Left in position, others can find them and learn from them but, hidden under your bed, they will be of little value to anyone! It is, however, sometimes worth while collecting fossils in working quarries or from coastal areas. In such places if you do not 'save' the fossil, it will probably be destroyed.

Always check loose rock for specimens rather than hacking away at the solid rock exposure. If you do find something, note it and make a sketch, because if you try to hammer it out you will probably just smash it.

Questions

For questions 1–5 write the letters (A, B, C, ...) of the correct answers in your notebook.

1 Which **two** of the following would be called trace fossils?
A shark's tooth
B stone axe
C dinosaur footprint
D echinoid spine
E marks left by worms burrowing in sediment

2 Which **one** of the following is **not** part of a typical trilobite?
A cephalon   B thorax   C nema
D pygidium   E pleura

3 Which **one** of the following statements is **correct**?
A Amphibians are thought to have evolved from ray–finned fish.
B Regular echinoids are thought to have evolved from irregular ones.
C Trilobites are used as zone fossils in Tertiary rocks.
D The earliest mammals evolved before the first birds.
E Marine plants evolved after land-living ones.

4 Which **one** of the following reptiles could fly?
A *Triceratops*   B *Tyrannosaurus*   C *Pterosaurus*
D *Brachiosaurus*   E *Plesiosaurus*

5 Which **three** of the following are extinct?
A trilobites   B gastropods   C graptolites
D ammonites   E brachiopods

6 Match the fossils in list 1 of the table below with the groups to which they belong (list 2) and a typical feature shown by them (list 3). For example, *Calymene* is a **trilobite** and has a **glabella**. Make a copy of the table in your notebook with lists 2 and 3 correctly arranged.

List 1	List 2	List 3
Calymene	coral	ambulacra
Micraster	bivalve	septal suture
Dibunophyllum	ammonite	pallial line
Calamites	echinoid	skull
Productus	plant	brachial valve
Carbonicola	trilobite	stem
Dactylioceras	hominid	glabella
Australopithecus	brachiopod	dissepiments

7 The diagram below illustrates the evolution of the vertebrates. Era A is the Cainozoic, period C is the Cambrian and line W represents the time range of the amphibians.

(a) Name era B.

(b) Name periods D, E and F.

(c) Name the vertebrate groups represented by lines V, X, Y and Z.

(d) During which period did the first birds evolve? Name a fossil example of one of the first birds.

(e) Write a short essay about 'The evolution of the first land-living vertebrates'.

8 Read the following paragraph.

On a recent fieldtrip I visited a quarry near the Yorkshire coast. I noted that the quarry face contained beds of fissile, fine grained rocks forming a large rounded anticline with limbs dipping 10° north and south. The manager told me that fossils were often found there and handed me a superb specimen of an ammonite. 'Tell me', he said, 'what sort of creature was this when it was alive?'

(a) Write a full answer to the quarry manager's question (be careful to explain the difference between the fossil and a living specimen: diagrams are essential).

(b) What age and type of rock was present in this quarry? Explain your answer?

(c) Draw a diagram to illustrate what the quarry face would look like.

9 The table below was made by a group of geologists after studying the fossils found in four beds of sedimentary rock. These beds are numbered 1 (oldest) to 4 (youngest). The ticks on the table show which fossils were found in each bed. For example, only fossils A, B and E were found in bed 1.

	Fossils					
	A	B	C	D	E	F
Bed 4		√	√			√
Bed 3		√	√	√		√
Bed 2	√	√			√	√
Bed 1	√	√			√	

(a) Which fossil comes from an organism which lived throughout the period of deposition of all four beds?

(b) Which two fossils became extinct at the same time?

(c) Which is the best zone fossil for bed 3?

(d) A geologist working in another area has discovered a bed of similar rock which contains fossils E and F; which of the beds in the table does this newly discovered bed correlate with?

(e) Fossil B is a colonial coral; what does this suggest about the environment in which beds 1 to 4 were deposited?

10 Study the map below which shows the boundary between two rock types. Note also the steep scarp slope which occurs along this boundary (see cross-section along line X–Y).

(a) Name fossils 1 and 2.

(b) Name the feature labelled S on each specimen

(c) What age is the shale containing fossil 1?

(d) What age is the shale containing fossil 2?

(e) What type of boundary do you think separates these shales: an unconformity, a fault or an ordinary bedding plane? (Explain your choice of answer.)

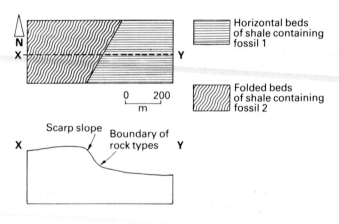

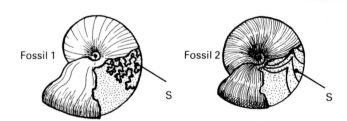

11 Write a short essay on 'What chance has a living organism of becoming a fossil?'

12 Arrange the following types of hominid into the correct order of evolution and comment on the brain capacity, tooth structure, diet and stance of each: *Ramapithecus / Homo erectus / Australopithecus / Dryopithecus*.

13 Plate tectonics

The basic ideas of plate movement were introduced earlier in this book. Now that you have a greater knowledge of rocks, structures and fossils we can study plate theory in more detail and see how well it can explain geological events and processes. Before continuing this chapter, re-read pages 13 to 17 to make sure you fully understand these introductory statements.

- The outer part of the earth is made up of a system of plates. Each plate is a solid slab of **lithosphere** about 100 km thick which includes the **crust** and part of the **upper mantle**.
- Plates move about the earth's surface as new material is added to some plate edges and re-moved (**subducted**) from others. As Fig. 2.21 showed, the movement can be likened to a giant 'conveyor belt' travelling at a few centimetres per year.
- A layer called the **asthenosphere** lies beneath the lithosphere and is thought to allow the movement of plates over the lower parts of the mantle.

- **Continents** are 'carried' slowly about the earth's surface because the plates beneath them are moving.
- Forces associated with plate movement cause intense geological activity to take place along the edges (**margins**) of plates. Plate margins can be divided into three main types, each of which has a distinctive set of surface features and underlying processes.

1. Constructive plate margins which occur where material is being added and plates are moving apart (See Fig. 2.19 on page 33, which shows the features of the Atlantic ocean floor)

2. Destructive plate margins which occur where material is being subducted and plates are moving together (see Fig. 2.20 on page 33, which shows the features of the Pacific ocean floor).

3. Conservative plate margins which occur where plates are sliding past one another with no material being added to or subducted from either side.

Fig. 13.1 The earth's plates

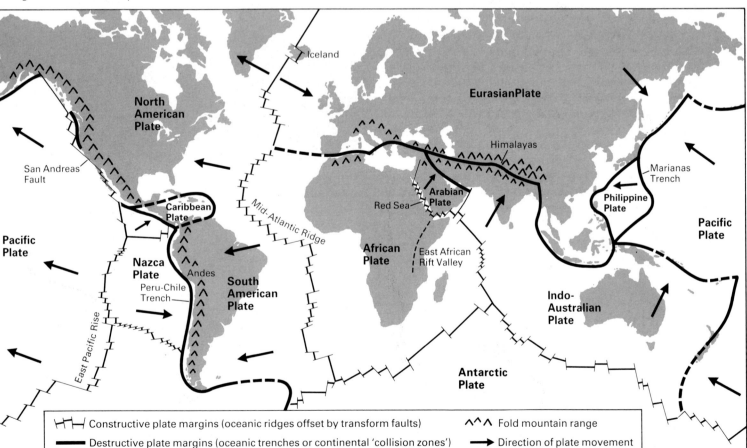

⊢⊣⊢⊣ Constructive plate margins (oceanic ridges offset by transform faults)	∧∧∧ Fold mountain range	
▬▬ Destructive plate margins (oceanic trenches or continental 'collision zones')	→ Direction of plate movement	

Constructive plate margins (spreading ridges)

These occur along **oceanic ridges** such as the East Pacific Rise and Mid-Atlantic Ridge. Two main processes take place to produce an effect known as **sea-floor spreading**.

1. New igneous material rises from the asthenosphere and is added to the plate margins bordering the ridge.

2. The plates on either side of the ridge move slowly outwards from it.

As yet, no one is sure which of these processes is a cause and which is an effect. It is possible that the addition of new material actually causes movement by pushing the plates apart. Alternatively the plates may be moving for some other reason (for example, because their opposite margins are being pulled by gravity into subduction zones) and this has the effect of creating a 'gap' which magmas rise and fill. A third, and more probable view, is that neither process causes the other because they are both working together as integral parts of a total system of plate movement.

Along the ridge itself, **basalt** lavas are erupted which cool rapidly in contact with the sea water and often show **pillow structures**. This basalt forms the topmost part of the new **oceanic crust** while, beneath it, additional magma cools more slowly to add deeper zones of coarser-grained denser igneous rock. In this way a complete new edge is constantly being added to plates as they spread outwards from a ridge. As Fig. 13.2 shows, it is not thought that the whole 100 km thickness of a plate develops immediately below a ridge. It seems more likely that thickness increases as movement away from the hot ridge region allows the **lithosphere** to cool.

The highest parts of an oceanic ridge may stand above sea level and form islands. For example, Iceland, the Azores and Tristan da Cunha are all parts of the Mid-Atlantic Ridge. Islands such as these have active volcanoes which erupt basalt lavas.

A programme of ocean floor surveys has provided the following evidence of sea-floor spreading.

1. Dating techniques show that ocean floor basalts, and the sediment lying immediately above them, get progressively older away from oceanic ridges. In fact the oldest parts of the oceanic crust are always found either in trenches above subduction zones or at the foot of continental slopes. Nowhere in the world have any such basalts been found which are more than 220 million years old so it seems that all our present system of ocean floors have developed since the Jurassic period. Of course, while these floors have been 'growing', other older floors must have been subducted and 'lost'.

2. It is known that every few hundred thousand years the earth's magnetism reverses its polarity. This

Fig. 13.3 Steam rising from a fractured basalt lava flow at Myratn, Iceland. Iceland is part of the Mid-Atlantic Ridge

Fig. 13.2 Constructive plate margin (sea-floor spreading)
Note: diagram exaggerates depth of oceans

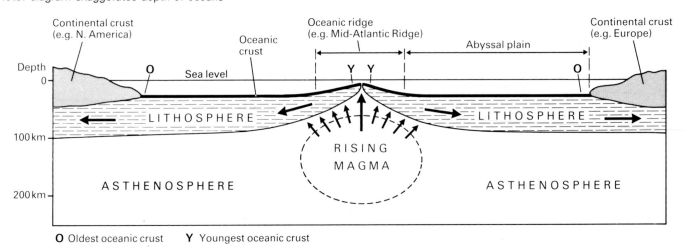

O Oldest oceanic crust   Y Youngest oceanic crust

means that if you had been living during certain geological times your compass would have pointed to the south pole instead of the north. Why such reversals happen need not concern us here, but their relevance to sea-floor spreading is important.

Ocean floor basalts contain a fairly high proportion of iron-rich minerals and, during crystallisation, particles of this iron become aligned by the magnetic field which is operating at the time. Because of this effect, a study of iron particle alignment in ancient basalts can be used to show when the earth's polarity was normal (that is, as it is at present) and when it was reversed. As Fig. 13.4 illustrates, surveys of ocean floor basalts have shown that a remarkably symmetrical pattern of normal and reversed magnetism exists on either side of an oceanic ridge. The only logical way such symmetry could have developed is for the ocean floor to have grown by steadily spreading outwards from the ridge; in doing so it would have recorded magnetic reversals in the way shown by Fig. 13.5.

3. It follows that, if oceans spread from central ridges, then the continents on either side must be moved apart. Further evidence of sea-floor spreading comes from studies which show that continents such as Africa and South America were once joined together. This will be considered in more detail in a later section on continental movement.

It also follows that, when an ocean first begins to open up and spread, a continent must be split into two parts by the newly developing ridge. This break-up will take place over a period of time and the continent will not fracture evenly along one straight line. As a result, the ridge which forms on the line of

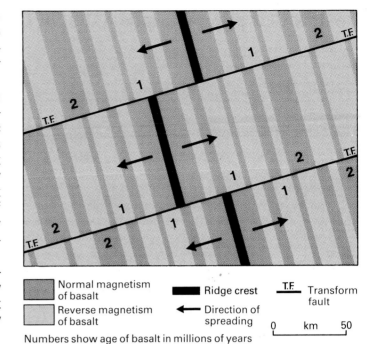

Normal magnetism of basalt
Reverse magnetism of basalt
Ridge crest
Direction of spreading
T.F. Transform fault
0 km 50
Numbers show age of basalt in millions of years

Fig. 13.4 Plan view of part of an oceanic ridge. Note the symmetrical pattern of normally and reversely magnetised basalt on either side of the ridge crest

fracture will not be straight either. Notice from Figs. 13.1 and 13.4 that the ridges actually have an offset pattern of distinct spreading units separated by **transform faults**.

By using the age and magnetic data explained above it is possible to calculate rates of spreading. This seems to vary between 2 cm and 10 cm per year for different parts of ridges in different oceans. It is an interesting thought that the Atlantic is growing at about the same rate as your fingernails!

Fig. 13.5 Formation of the magnetic reversal pattern shown by sea-floor basalts (Reversals occur every 200 000 to 600 000 years)

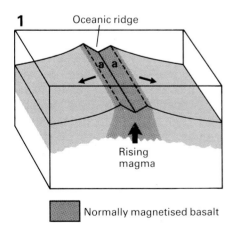

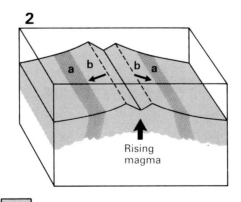

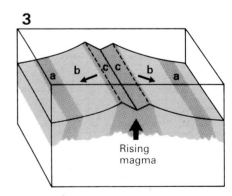

Normally magnetised basalt   Reversely magnetised basalt

Stage 1 Spreading produces a zone of normally magnetised basalt (labelled a)

Stage 2 Polarity of the earth changes. Basalts now forming (labelled b) show reversed magnetism. Note how the original normally magnetised basalt (labelled a) becomes split into two equal 'stripes' by the arrival of the newer material.

Stage 3 Polarity changes back, and normally magnetised basalt (labelled c) is produced once more. Note how a symmetrical pattern of basalt magnetism is developing.

Destructive plate margins (subduction zones)

These occur at **ocean trenches** such as the Marianas Trench and the Peru–Chile Trench. They are places where plates are moving towards each other, forcing one to descend beneath the edge of the other into a **subduction zone** (see Fig. 13.6). You should remember the following points about subduction zones.

● Evidence of their size and shape was collected by a geologist called Hugo Benioff. He plotted the depth and position of earthquake foci near ocean trenches and discovered that activity occurred in a distinct zone dipping at about 45° into the asthenosphere. Because of his work, subduction zones are often called **Benioff zones**.

● Since plate material is originally formed (at ridges) by magmas rising from the asthenosphere, at subduction zones it is in a sense returning to the level it came from. During subduction, the upper layer of **oceanic (basaltic) crust** is partly melted and absorbed at depths of 100–300 km but the rest of the **lithosphere** continues down to about 700 km before being completely 'digested'.

● Melting takes place in subduction zones because of increasing depth and the effects of friction. Part of the melting process produces magmas which rise back up towards the surface. These magmas tend to be far more varied in composition than the monotonous basalts of ridge areas, and a wide range of igneous activity can occur at destructive plate margins.

● Only oceanic plate material can be subducted; the **granitic crust** of continents is not dense enough to descend into the asthenosphere.

● Because the plates are moving towards each other, destructive margins are under constant powerful forces of compression (that is, being squeezed from both sides). As a result, many of these regions show the effects of **folding** and **regional metamorphism**: features that rarely occur under the tensional forces acting at constructive margins.

● As mentioned earlier in this chapter, geologists are unsure whether plates are 'pushed' down subduction zones by the action of sea-floor spreading or 'pulled' down by gravity. Perhaps both effects work together?

● Although subduction always involves the descent of oceanic plate material at deep ocean trenches, different situations can occur depending on the position of nearby continents. There are three possible types of destructive plate margins:
(a) **oceanic plate meets oceanic plate** (no continent involved);
(b) **oceanic plate meets continental edge**;
(c) **collision of two continental edges** as a result of the oceanic plate material between them being completely subducted.

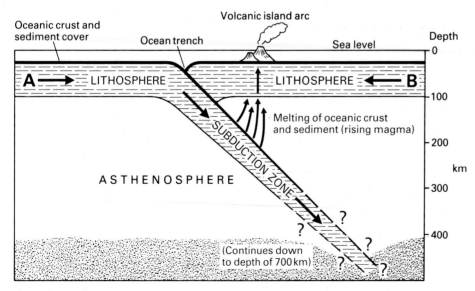

Fig. 13.6 Destructive plate margin (oceanic plate meets oceanic plate)
Note: diagram exaggerates depth of ocean

Oceanic plate meets oceanic plate (see Fig. 13.6)

The most distinctive feature to develop in this situation is a **volcanic island arc** (a string of small islands), formed by magmas rising from the subduction zone below. Since the descending plate contains oceanic crust, we might expect this magma to be basaltic. However, the volcanoes of island arcs usually erupt **andesite**. There are several possible theories for the origin of this andesite but, before reading them, refer back to Fig. 5.19 on page 40 and remind yourself of the difference in composition between basalt and andesite (andesite is less dense than basalt and contains a higher proportion of silicon and oxygen).

Andesite is possibly formed at subduction zones because the basaltic material has been 'contaminated' by silica-rich sediments. A thin covering of clays and oozes (ocean-floor deposits of the **abyssal plain**: see page 58) will have collected on the surface of the oceanic crust by the time it reaches a subduction zone. Although most of this will be scraped off and deformed in the trench, some will be carried down with the descending plate. At depths of 100–200 km the basaltic oceanic crust, and its remaining sedimentary cover, will begin to melt and mix. This produces a magma richer in silica than normal basalt would be (that is, magma with the composition of andesite). Note that the 'extra' silicon and oxygen has come from the melting of sedimentary particles such as clay minerals.

It is also possible that some andesite is produced by **fractionation** of rising basalt magma. In this process, most of the magma crystallises before reaching the surface and leaves only the least dense part (fraction) to continue upwards and erupt. Since this fraction contains a higher proportion of lighter elements such as silicon and oxygen it produces andesite.

'Oceanic to oceanic' subduction zones are common around the Pacific. Look at Fig. 13.7 and use an atlas to name some of the trenches and island arcs shown.

Before moving on, let us consider one very important point about all types of destructive plate margins. Many people find it hard to understand how it is possible for both plates to be moving towards one another (note arrows A and B in Fig. 13.6) when only one plate is being subducted. The easiest way to illustrate how this happens is to make a 'working model' with two sheets of paper representing the

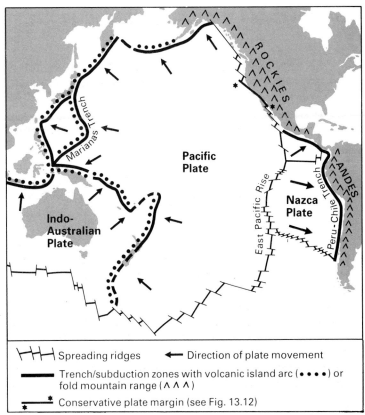

Fig. 13.7 Plate margins around the Pacific

plates. Fig. 13.8 explains how to do this. When you have tried the demonstration for yourself check that you understand the principles of movement by answering this question.

If, in Fig. 13.6, arrow A represents movement of 4 cm/year *and arrow B movement of* 3 cm/year, *what is* (i) *the total rate of subduction* (ii) *the rate and direction of movement of the site of the ocean trench?*

Fig. 13.8 Simple model to demonstrate the relative movement of two plates at a destructive margin

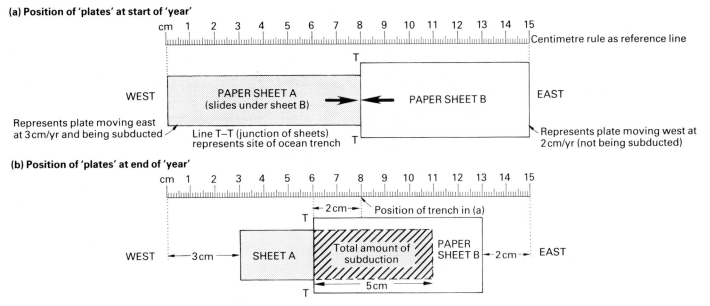

Note how the site of the ocean trench has moved in the same direction and at the same rate as the plate represented by sheet B.

Oceanic plate meets continental edge
(see Fig. 13.9)

In this situation it is always the oceanic plate which is subducted; continental crust is not dense enough to descend into the asthenosphere.

To understand this type of destructive margin we first need to consider what is likely to have happened to a continent before subduction began to affect its edge. This edge would have originally been under the sea and, over millions of years, its continental shelf and slope would have received large amounts of sediment. In fact, such offshore areas tend to subside and troughs of material up to 5 km thick can sometimes develop here. These troughs typically contain deep-water **shales** with occasional beds of coarser **greywacke sandstone** swept in by **turbidity currents** (see page 76).

Once subduction begins, the accumulated thickness of offshore deposits becomes 'caught in the vice' of plate movement and an **orogeny** takes place. Compression of the shelf/slope region causes the whole sequence of sediments to deform, rise and develop into a range of **fold mountains**. Such a range will extend for thousands of kilometres, right along the continental edge. An example of this is the Andes Mountains which have been formed by subduction of the Nazca Plate beneath the western edge of the South American Plate. On the highest Andean peaks, marine fossils can be found, showing that these mountains originally began as sediments on the sea bed of the continental shelf and slope. Because the whole Andean Range has been produced by forces of compression acting from east and west, the axes of its folds run from north to south.

During an orogeny, the deeply buried regions will undergo high-grade **regional metamorphism**. A sequence of rocks like that shown in Fig. 8.11 (page 85) may develop, with a massive body of **granite** grading outwards into surrounding zones of **gneiss** and **schist**. Further from the metamorphic centre, compression produces **folding** and causes the development of **cleavage** in the rocks. Folding is frequently so intense that **nappe folds** and **thrust faults** develop.

Volcanic action will also take place in the rising mountain range. As with island arcs, the most commonly erupted lava is **andesite** (this rock is named after a place where it is likely to be found: the Andes Mountains). However, because more (silica-rich) sediment is supplied into trenches along continental edges and because magma has to rise through the silica-rich (granitic) crust of the continent to reach the mountain chain, eruptions of **rhyolite** are also likely to occur. Rhyolite lava contains an even higher proportion of silica than andesite or basalt (see Fig. 5.19 on page 40).

Towards the end of an orogeny, when subduction stops, the mountain range may adjust to the release of pressure. Some regions will then be affected by **normal faulting** and **horsts** or **rift valleys** can occur (see Fig. 10.17, page 99). The release of pressure also allows any remaining magmas to rise towards higher levels where they intrude into folded sediments as relatively small **batholiths** with distinct **metamorphic aureoles**.

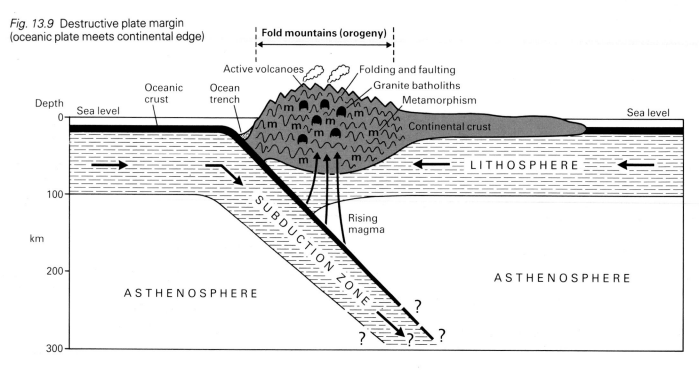

Fig. 13.9 Destructive plate margin (oceanic plate meets continental edge)

Continental collision

(see Figs 13.10 and 13.11)

Continental collision takes place when a complete ocean is closed by subduction. Plate movements of this type cause **orogeny** with folding, metamorphism, intrusions and volcanics similar to those explained in the previous section.

Fig. 13.11 illustrates the developments on a collision zone. As you study the diagrams take note of the following points.

1. Before continental collision actually takes place a mountain range will already have developed on the edge of the continent directly above the subduction zone (Fig. 13.10a).

2. The continent carried on the subducting plate cannot descend into the asthenosphere (granitic crust is not dense enough). Its edge must therefore begin buckling and uplifting as it approaches the point of subduction (Fig. 13.10a).

3. Eventually the two continents unite as a single landmass and their deformed edges become merged into one great mountain range. At this point, subduction must stop because there is no ocean left. However, magmas still rise, and forces affect the region for a considerable time after the ocean has closed. This is because material already in the subduction zone is continuing to descend (Fig. 13.10b).

The best-known example of orogeny caused by continental collision is the development of the **Himalayas**. This occurred in the Tertiary period when an oceanic region of the Indian Plate was totally subducted beneath the Eurasian Plate.

There are a number of different possibilities for continental collision. Sometimes, both continents may have subduction zones dipping beneath them so that the closing of the ocean takes place as shown in Fig 13.11. To make matters even more complicated, island arcs may sometimes be carried into collision zones and involved in the resulting orogeny.

Fig. 13.10 Destructive plate margin: continental collision

(a) Collision begins

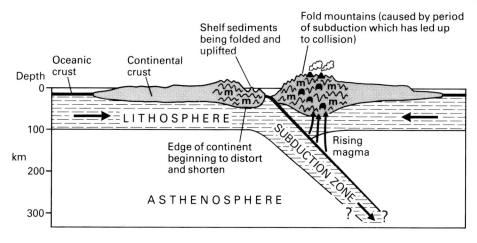

(b) Collision complete

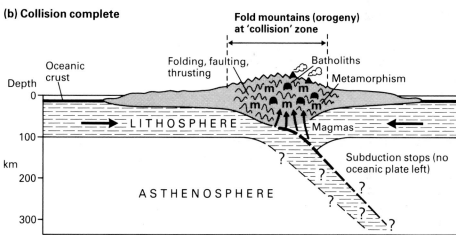

Fig. 13.11 Continental collision with subduction on both sides of ocean

(a)

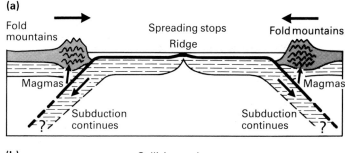

(b)

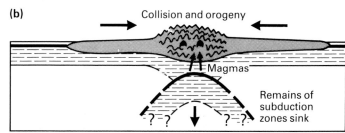

145

Conservative plate margins

These are margins where one plate slides past another with nothing added or taken away from either edge. An example occurs where the North American Plate comes into contact with the spreading ridge of the East Pacific Rise. As shown by the map in Fig. 13.12, the junction is marked by a **transform fault** called the **San Andreas**. Because friction prevents the plate edges from sliding smoothly, movement tends to occur as a series of sudden jerks, each accompanied by an earthquake (would you like to live in one of San Francisco's high-rise buildings?). Another interesting point is that the Californian coast and peninsula are moving towards Alaska at an average rate of 5 cm per year!

Fig. 13.12 Conservative plate margin: San Andreas Fault

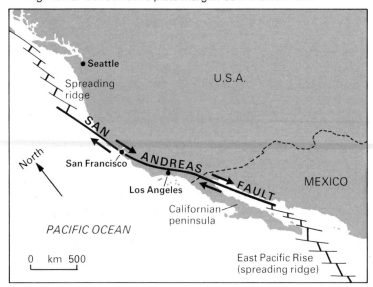

Fig. 13.13 Aerial view of the San Andreas Fault. Note how the stream has been offset by about 400 m along the fault line (F–F)

Continental break-up and movement

If sea-floor spreading takes place it follows that continents must break up and move about the earth's surface. There is considerable evidence for continental movement and we shall illustrate some of it by considering whether Africa could once have been joined to South America. Refer to Fig. 13.14 as you consider these points.

● Even on a normal map the coastlines of these continents look as though they would fit together like pieces of a jigsaw. If the maps are redrawn to show the true edge of the continents (that is, at the foot of their continental slopes) then the fit is even more convincing.

● When the continents are placed together their geology matches remarkably well and boundaries between different rock units line up across the junction. Fig. 13.14 shows the situation for Pre-Cambrian rocks but there are further similarities if other rocks are plotted.

● Fossils of *Mesosaurus* are found in both Africa and South America. These small reptiles are known to have lived in fresh-water lakes and it seems extremely unlikely that they could ever have crossed a wide ocean.

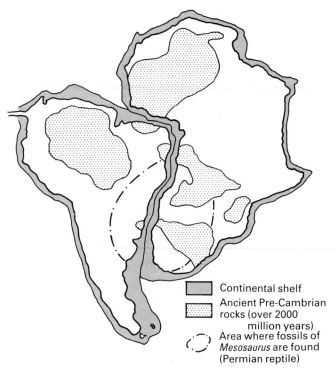

Fig. 13.14 Evidence for continental movement: South America and Africa

• Evidence of continental movement also comes from the distribution of sedimentary rocks that could only have formed in certain climates. Very similar glacial deposits of Upper Carboniferous age occur in both South America and Africa but, as Fig. 13.15 shows, these particular rocks are also found on other landmasses and allow an even bigger 'continental jigsaw' to be put together. It seems that, during the Upper Carboniferous, all the southern continents were together as a supercontinent which geologists call **Gondwanaland**.

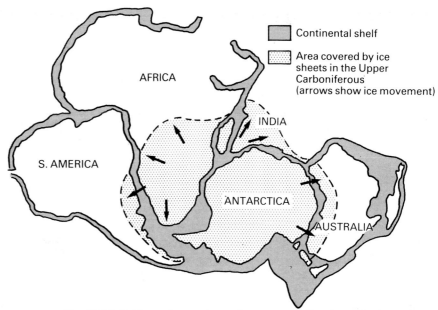

Continental shelf

Area covered by ice sheets in the Upper Carboniferous (arrows show ice movement)

Fig. 13.15 Evidence for continental movement: Gondwanaland

Gondwanaland was not the largest continent to have existed in the past. There is evidence to show that in Lower Permian times Europe, North America and Asia also became joined to it and, remarkable as it may sound, all the world's continents then formed one vast single landmass known as **Pangaea**. Pangaea broke up during the Mesozoic era as our present system of oceans began to open and spread. Some of the main aspects of this break-up are summarised in the lower part of Fig. 13.18.

Exactly how continental break-up occurs is not fully understood but certainly the process must be associated with the early developments of a spreading ridge. It is possible that parts of Africa are beginning to separate at present. Notice, from Fig. 13.16, how the area of the **Red Sea** is rifting as a result of tension. The rifting has allowed a section of continental crust to subside and provides a 'weak point' for magma to reach the surface. This is basaltic material so it seems likely that it may represent the beginnings of a new ridge system which, eventually, will cause an ocean to spread between Africa and Arabia. The nearby **East African Rift Valley** may represent a younger stage of break-up. Here there is rifting but as yet only limited volcanic activity is taking place and the sea has not flooded in. An alternative theory, supported by many geologists, is that this rift valley is actually the site of a potential spreading ridge which has failed to break through.

Fig. 13.16 Continental break-up: Africa

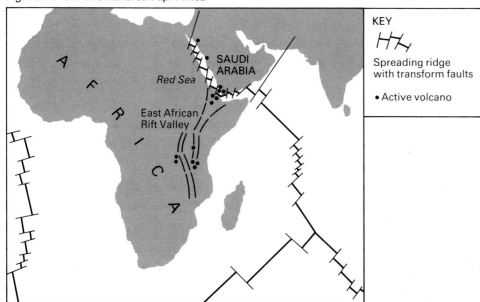

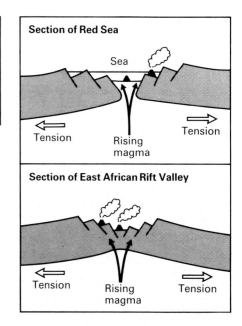

KEY

Spreading ridge with transform faults

• Active volcano

Section of Red Sea

Sea

Tension   Rising magma   Tension

Section of East African Rift Valley

Tension   Rising magma   Tension

Explanation of plate movement

The simplest suggested explanation is the **convection current theory**. It is based on the way liquids are known to behave when heated from below. Look at Fig. 13.17a; the heated liquid begins to rise because it has expanded and become slightly less dense. It then spreads out across the surface, cools, and sinks back down the sides to be reheated and made to rise again. The overall circular movement is called a **convection cell**. *It can be easily demonstrated by slowly heating a pan of water over a bunsen flame. Add a few crystals of potassium permanganate to the centre. Their coloration shows how the water is moving.*

Fig. 13.17b shows the theory of convection in the asthenosphere and deeper parts of the mantle. Heat from the earth's interior is said to cause rising convection currents which supply material to spreading ridges. Subduction zones are seen as the cooling and descending parts of the cells.

Fig. 13.17 Convection currents and plate tectonics

(a) Convection cells in a pan

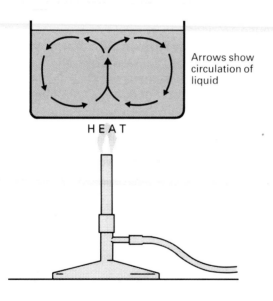

Arrows show circulation of liquid

HEAT

(b) Convection cells in the earth's mantle?

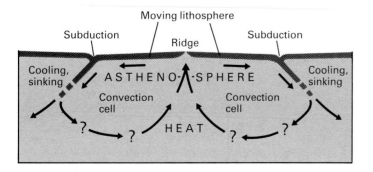

There is a lot to be said in favour of the convection current theory of plate tectonics. In particular, it provides a complete system, with surface movements of the lithosphere being balanced by a return flow at depth. However, as the following arguments show, simple convection is unlikely to be the total answer.

1. If convection currents were operating as a uniform system of circular cells we would expect a system of alternating ridges and subduction zones around the world. This is not the case: around the equator, for example, moving westwards from Africa the sequence is ridge, subduction, ridge, subduction, subduction, ridge.

2. There are some localised hot spots where heat rises well away from spreading ridges: for example the Hawaiian volcanic islands in the centre of the Pacific Plate. How can convection cells explain this?

3. Information from earthquake waves suggests that much of the mantle is in too rigid a state to move like a liquid under convection.

Just as geologists had to wait many years for the evidence of plate movement to be discovered, it seems we must wait a little longer before understanding exactly what makes the system work.

Plate tectonics and sea level

Plate tectonics links most aspects of geology together and provides many of the answers about why, how, when and where different geological processes take place. It may explain even more than continental movement, mountain building and the formation of igneous, sedimentary and metamorphic rocks. *Look at Fig. 13.18. What is the relationship between the arrangement of past continents and world sea level?*

World sea level was at its highest during the Upper Cretaceous period, and it is interesting that this was also the time when Gondwanaland and Laurasia (a supercontinent consisting of North America, Europe and Asia) were splitting and moving apart at an increased rate. Continental break-up occurs because of spreading ridges, and these take up a lot of space under the sea. (Our present system of ridges is reckoned to occupy about 160 million km^3 of the space available in ocean basins.) It therefore seems logical that, at times of rapid sea-floor spreading, the world sea level must rise as water is displaced by the many ridge systems.

Plate movement is not the only possible cause of sea-level changes. For example, during periods of glaciation, sea level may be lowered because a large amount of water is 'trapped' on land in the form of ice. This is the situation at present and was also a factor in the particularly low sea levels which occurred during the late Palaeozoic era when ice sheets covered much of Gondwanaland.

Fig. 13.18 Plate tectonics and sea level

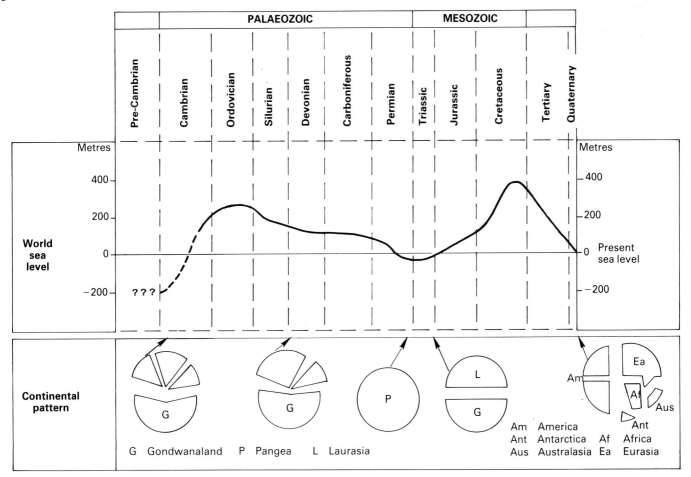

Table 13.1 Plate margins

Type of margin	Surface features	Effects of earth forces	Igneous activity	Metamorphism	Examples
Constructive (sea-floor spreading)	Oceanic ridge	Earthquakes (shallow focus) Transform faults	Basalt lavas (with dolerite and gabbro intrusions beneath)	—	Mid-Atlantic Ridge East Pacific Rise
Destructive (subduction)					
(a) Oceanic plate meets another oceanic plate	Ocean trench Island arc	Earthquakes (deep focus)	Andesite lavas form volcanic islands	—	Philippine Islands Aleutian Islands
(b) Oceanic plate meets edge of continent	Ocean trench Fold mountain range on edge of continent (i.e. an **orogeny**)	Earthquakes (depth variable) Folding, faulting and uplift of continental edge	Andesite and rhyolite lavas Large batholiths of granite in mountain range	Regional metamorphism in depths of mountain range	Andes Mountains
(c) Two continental edges meet and collide	Fold mountains along line of collision (i.e. an **orogeny**)	Earthquakes (depth variable) Folding, faulting and uplift	Andesite and rhyolite lavas Large batholiths of granite in mountain range	Regional metamorphism in depths of mountain range	Himalayan Mountains
Conservative (plates 'slide' past each other)	'Fault zone'	Earthquakes along transform fault	—	—	San Andreas Fault (California)

Questions

For questions 1–5 write the letters (A, B, C, . . .) of the correct answers in your notebook.

1 The most common type of igneous rock to be formed at a constructive plate margin is
  A rhyolite   *B* granite   *C* basalt
  D obsidian   *E* andesite

2 The average rate of plate movement is
  A 0.1 to 0.5 cm/year   *B* 2 to 10 cm/year
  C 10 to 20 cm/year   *D* 25 to 50 cm/year
  E over 100 cm/year

3 Which **one** of the following lists correctly names the continents which once formed Gondwanaland?
  A North America, South America and Europe
  B Europe, Asia and Australia
  C Africa, Asia, India and North America
  D Africa, South America, Australia, India and Antarctica
  E Europe, Iceland, India and Asia

4 Which **one** of the following plates spreads from the East Pacific Rise and descends into a subduction zone at the Peru-Chile Trench?
  A Nazca Plate   *B* Eurasian Plate   *C* Pacific Plate
  D South American Plate   *E* Indo-Australian Plate

5 During which **one** of the following periods were all the world's continents joined together as Pangaea?
  A Tertiary   *B* Ordovician   *C* Devonian
  D Permian   *E* Cretaceous

6 Name the type of plate margins which occur in these places:
  (a) the Aleutian Islands
  (b) Iceland
  (c) the Caribbean
  (d) the Red Sea
  (e) the Mediterranean
  (f) the East African Rift Valley
  (g) southern California
  (h) the Philippines

7 Look at the diagram below which shows a cross-section through several plates.

(a) Name features A, B and C.
(b) What type of crust is shown by the dotted shading?
(c) State the directions you would expect plates Y and Z to be moving in.
(d) What name is given to the zone beneath the lithosphere (labelled D on the diagram); how is this zone thought to aid plate movement?
(e) Explain, as fully as possible, what types of geological processes are involved in the formation of the fold mountain range
(f) Explain what is happening beneath feature A; how are events here related to the movements of plates X and Y?

8 Study the graph below which shows data collected by a survey across an ocean floor.
(a) How old is the ocean floor basalt at:
   (i) the edges of the ocean,
   (ii) the centre of the ocean,
   (iii) 400 km from the centre of the ocean?
(b) Carefully explain how the evidence shown by this graph supports the theory of sea-floor spreading; what type of plate margin would be present in the centre of this ocean?
(c) What results would you expect if another survey was carried out to measure the magnetism of the basalts across this ocean floor?

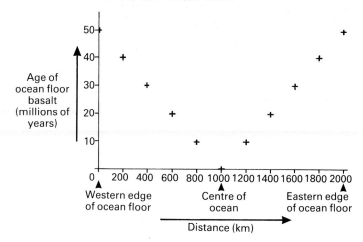

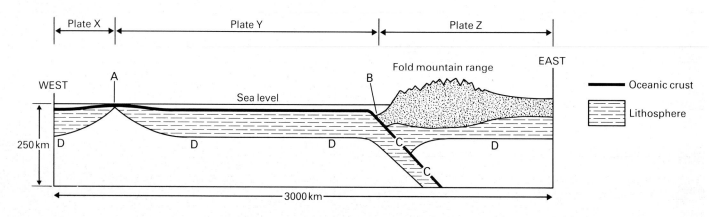

150

14 The geological history of the British Isles

The oldest rocks in Britain were formed over 2900 million years ago; the youngest are still being deposited today. The long story of Britain's geological development has been worked out by the following.

- Studying the mineralogy, texture, structure and fossil content of rocks. This gives information about how and where they were formed and what has happened to them since formation.
- Applying dating methods to rocks and the events which have affected them. This may involve **relative dating** (see pages 110 and 118) or **absolute (radiometric) dating** (see page 90).
- Correlating information from different areas. This gives a better understanding of past events by showing the possible variation and extent of past environments.

With so much time and evidence to consider, Britain's geological history can appear to be a huge list of seemingly separate events. However, you should remember that geological events are not completely separate but follow each other in a continuous succession. To recognise the most important trends of this succession we shall study Britain's past in relation to **plate movement**. This can provide possible answers to questions such as the following.

1. Why has Britain had different climates during different periods of time (as shown from the evidence of sedimentary rock sequences and fossils)? This can be explained quite logically if we think of how plate tectonics causes continents to move about the surface of the planet. There is considerable evidence to suggest that, since the Cambrian period, the British part of the European plate system has moved steadily northwards from approximately 40° south of the equator to its present position of 55° north. This makes it easy to see why a sequence of different climates and a sequence of different rock types occurred. For example, it is thought that the forests of the Upper Carboniferous (which eventually became coal deposits) grew particularly well at that time because Britain was then close to the equator. Similarly, Britain's Permian and Triassic red sandstones and evaporites developed because, by that time, Britain had moved north from the equator into desert conditions (that is, it was in the same latitudes as the Sahara is now).

2. Why have mountain building, igneous activity and regional metamorphism only occurred in Britain at certain specific times? It seems logical to suggest that these events happened when Britain was close to an active plate margin.

3. Why have major changes in sea level taken place? It is likely that some of these resulted from changes in the rate of development of spreading ridges, as explained on page 148.

Fig. 14.1 Silurian rocks showing effects of the Caledonian Orogeny, Shap Fell, Cumbria

To explain Britain's geological history in terms of plate movement we shall divide the story into five main episodes; each separated from the next by the most dramatic effect of plate motion: an **orogeny**. Look at Fig. 14.2 which illustrates these episodes and the orogenies associated with them. Note that the episodes vary in length and the orogenies vary in the amount of deformation they caused. Rocks from each episode also tend to be separated by **unconformities**: this is because the uplift of each orogeny meant that erosion generally took place before the next series of sediments was deposited.

Fig. 14.3 shows the solid (bed-rock) geology of the British Isles. Notice how rocks from the oldest episodes tend to occur in the north and west while those from later episodes are mainly found in the south and east. Notice too that because the map has been drawn to show the solid geology, materials from episode 5 have had to be left out. These recent materials have not yet been lithified into solid rock but form a surface covering over much of the lowland and practically all of the sea floor.

Before studying each episode more closely, several other points must be made.

- The further back in time we go, the less information we have. The story of the older episodes is therefore less precise than that of more recent ones.
- The present geographical shape of Britain has only existed for about 10 000 years (and it is still slowly changing). When we talk of 'Britain in the past' we really mean 'the area which eventually became Britain'.
- In the following pages the letters Ma are used as shorthand for 'million years ago'.

Fig. 14.2 The major episodes in Britain's geological history
Note: (a) the height of each block shows age range *not* thickness of rock involved; (b) blocks are 'stepped' to show that some parts of each episode are *exposed* (E) while others remain *hidden* (H) beneath rocks from the later episodes

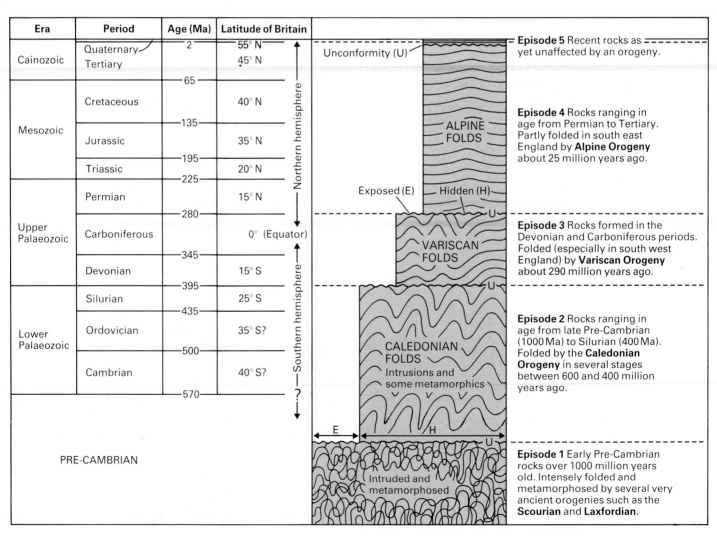

Fig. 14.3 Rock outcrops of major episodes

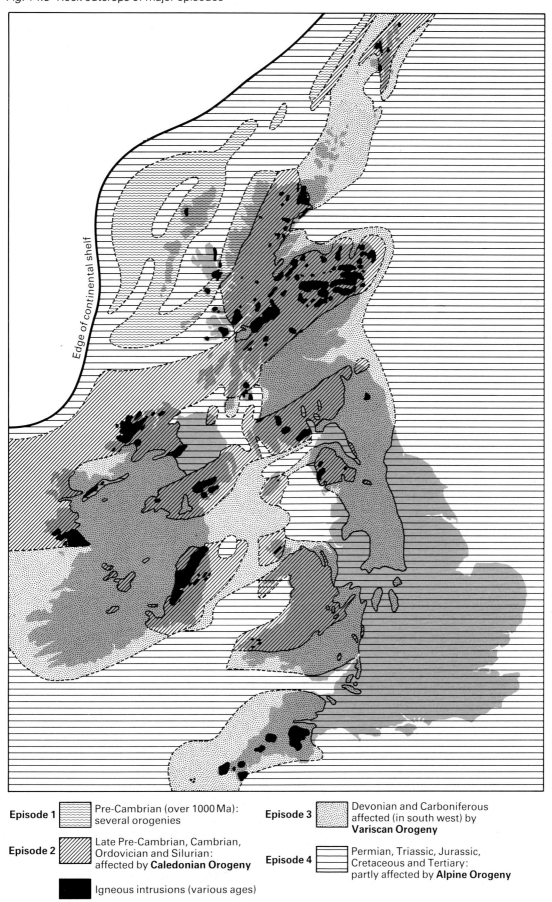

Episode 1 Pre-Cambrian (over 1000 Ma): several orogenies

Episode 2 Late Pre-Cambrian, Cambrian, Ordovician and Silurian: affected by **Caledonian Orogeny**

Igneous intrusions (various ages)

Episode 3 Devonian and Carboniferous affected (in south west) by **Variscan Orogeny**

Episode 4 Permian, Triassic, Jurassic, Cretaceous and Tertiary: partly affected by **Alpine Orogeny**

Episode 1: the earliest evidence

(see Fig. 14.4)

Here we deal with events which took place over 1000 million years ago. It is the largest but least understood episode of Britain's geological history. A look back to page 89 will remind you of the difficulties involved in the interpretation of such ancient Pre-Cambrian rocks. Nearly all the evidence has by now been lost, hidden or altered by erosion, burial and metamorphism.

Britain's oldest rocks are the **Lewisian Gneisses** exposed in the Hebrides and along Scotland's north west coast. Studies show that their high-grade regional metamorphism was produced during two separate orogenies. However, there is simply not enough evidence to suggest what plate movements or continental collisions could have been responsible. The orogenies are known as the **Scourian** (about 2700 Ma) and the **Laxfordian** (about 1800 Ma), but even they do not give the real beginning to our story. Other, even older, rocks must have existed before this to provide the eroded material which formed the sediments which became altered to gneiss during the Scourian Orogeny!

Apart from the Lewisian Gneisses, the only other rocks over 1000 Ma to be seen in Britain are some small areas of metamorphics in the Channel Islands and south east Ireland. However, it is likely that other equally ancient rocks could be found beneath much of the country if boreholes were drilled right through the thick sequences of later rocks which lie above them.

Fig. 14.4 Distribution of rocks over 1000 million years old (the Lewisian Gneisses)

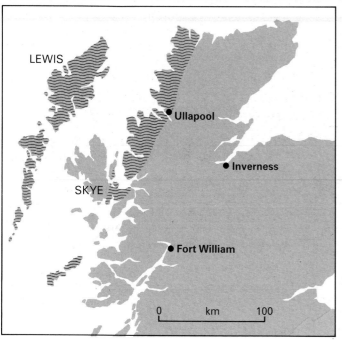

Episode 2: a closing ocean brings Britain together

Although this is the second episode, it is the first part of the story that can be interpreted in a meaningful way. It covers events during the late Pre-Cambrian and the Lower Palaeozoic (that is between 1000 and 400 million years ago). Rocks of this age are widely exposed in the highlands of north and west Britain (see Figs. 14.3 and 14.13) and all show the effects of the **Caledonian Orogeny**.

Before studying any details of this episode you should look at Fig. 14.5a. On this map Europe and North America are shown without the Atlantic Ocean in between. Although this may look strange, you have to realise that the Atlantic did not open and spread until the Mesozoic era (see page 147). Without separation by the Atlantic, some important links become clear.

- Similar late Pre-Cambrian and Lower Palaeozoic rocks in Scandinavia, Greenland and Canada join with those in Britain to form a continuous belt with the same general direction of folding. Geologists call this the **Caledonian Fold Belt**.
- The fold belt is sandwiched between two areas of even older rocks.

This pattern suggests that Caledonian folding was produced by the collision of two ancient continents. However, as you would expect with events covering 600 million years, the story is open to several different interpretations. The following pages outline the possible stages, but as you read them keep the following points in mind.

Fig. 14.5 The Caledonian Fold Belt

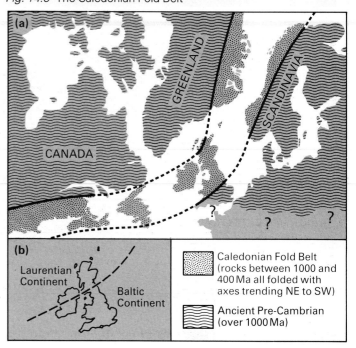

1. The continents involved were basically those shown on either side of the fold belt (see Fig. 14.5b). We shall call them:

(a) the **Laurentian Continent** (the remains of this now form the ancient Pre-Cambrian regions of Canada and Greenland);

(b) the **Baltic Continent** (the remains of this now form the ancient Pre-Cambrian regions of Northern Europe).

2. During the late Pre-Cambrian and the Lower Palaeozoic, these continents were separated by an ocean known as **Iapetus** (remember this was *not* the Atlantic but a different, much earlier, ocean). Iapetus also separated Britain into two parts because, at this time, the area which was eventually to become Scotland and northern Ireland formed part of the Laurentian Continent while England, Wales and southern Ireland were part of the Baltic Continent. Britain was 'brought together' when these continents collided during the late Silurian. It is an interesting thought that the present day border between England and Scotland runs approximately along the junction of what were once parts of separate continents!

3. Palaeomagnetic evidence suggests that Iapetus and its surrounding continents were in the southern hemisphere.

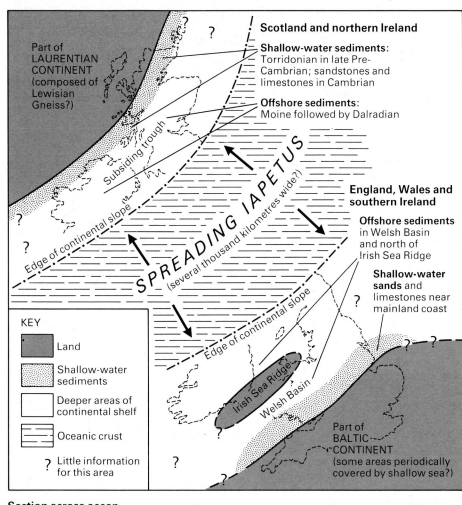

Fig. 14.6 Iapetus in the Cambrian period (some aspects of late Pre-Cambrian are also shown)

Iapetus in the late Pre-Cambrian and Cambrian

(1000 to 500 Ma : Fig. 14.6)

Since Britain was then divided by the Iapetus ocean it is easier to consider its geology in two separate parts.

Scotland and northern Ireland

This area lay on the edge of the Laurentian Continent where, about 1000 million years ago, sediments began to be deposited on a landscape of eroded Lewisian Gneisses. As Fig. 14.6 shows, part of the region was then a shoreline and part on the continental shelf. The shelf became the site of a subsiding trough which, for over 400 million years, continued to be slowly infilled by fine-grained sediments. The great thickness of rocks formed in this offshore trough has been divided into two groups. The older lower part is entirely of Pre-Cambrian age and known as the **Moine Group**. The upper part is called the **Dalradian Group** and is partly Pre-Cambrian and partly Cambrian in age. Moine and Dalradian rocks now form the rugged landscape of the Scottish Highlands but they cannot be studied in their original sedimentary state. As you will discover by reading on, these rocks were highly metamorphosed during the Ordovician.

Fig. 14.7 Torridonian Sandstone lying unconformably on Lewisian Gneiss at Stac Pollaidh, Scottish Highlands

Fig. 14.8 Welsh slate quarry

While the prolonged deposition of offshore sediments was taking place in the trough, a much thinner sequence of shallow-water sediments was forming closer to the Laurentian shoreline. During the late Pre-Cambrian, deltas produced **Torridonian Sandstones** but changes in sea level meant that some of these were eroded before limestones and orthoquartzites were deposited in the Cambrian.

England/Wales/southern Ireland

This area lay on the edge of the Baltic Continent and during the late Pre-Cambrian another trough had been filling and subsiding on this continental shelf. About 625 Ma, the area was affected by subduction which metamorphosed and uplifted the trough sediments into a narrow chain of intensely deformed mountains. The eroded remains of this chain can be seen in Pre-Cambrian rocks of Anglesey and south east Ireland.

Since there are so few outcrops of Pre-Cambrian rocks in England it is difficult to tell what was happening close to the shores of the Baltic Continent at this time. Evidence from the small exposures that do occur (for example at Church Stretton in Shropshire) suggests that a shallow sea covered the area and that volcanic eruptions were widespread. However, by about 600 Ma, most of these rocks had been folded. Presumably the volcanics and folding were related to the subduction taking place further offshore.

By the beginning of the Cambrian period, subduction along the edge of the Baltic Continent seems to have ended. At about this time sea level rose (indicating an increased rate of spreading by Iapetus?) and the main fold belt of Pre-Cambrian rocks then stood as an offshore ridge. This feature (known as the **Irish Sea Ridge**: see Fig. 14.6) supplied sediment to the sea on either side. To its north, shales were deposited on the edge of the continental shelf (these rocks now lie buried beneath Ordovician material in such places as the Isle of Man and the Lake District). To the south side of the Irish Sea Ridge, Cambrian sediments accumulated in a region of deepening water called the **Welsh Basin**. Rocks from this environment are now seen in north west Wales where they include some excellent roofing slates (see Fig. 14.8 but note that these rocks were not metamorphosed from the original shale until Iapetus closed in the late Silurian).

Across the south east of the Welsh Basin, close to the shore of the Baltic Continent, was a region of much shallower sea. During the Cambrian period thin beds of sandstones and limestones were deposited here to lie unconformably on eroded late Pre-Cambrian sediments and volcanics.

A note on Cambrian fossils

The Cambrian is the first period when fossils are found frequently enough to be used for correlation. Trilobites are most common and, in addition to their value as zone fossils, they add further evidence that Iapetus existed. Two different sets of Cambrian species are found: one occurs in the rocks of Scotland and Canada while the other is preserved in the rocks of England, Wales and southern Scandinavia. This separation of species would seem to suggest that the two regions were then separate continents with a wide enough ocean between them to prevent the migration of sea-floor creatures such as trilobites.

Iapetus in the Ordovician

(500 to 435 Ma: Fig. 14.9)

Although Britain remained in two separate parts during the Ordovician, Iapetus was, by then, a closing ocean. This period is easier to correlate than earlier times: graptolites provide good zone fossils for fine-grained offshore sediments while trilobites and brachiopods are used for deposits from shallower waters.

Scotland and northern Ireland

Events here were dominated by mountain building as the northern edge of Iapetus began to subduct beneath the Laurentian Continent. **Moine** and **Dalradian** sediments (which had been lying in the offshore trough) were intensely folded and metamorphosed as the whole area of the shelf and slope was compressed and uplifted. Compression was so strong that some Moine rocks were pushed many kilometres towards the north west along the **Moine Thrust** (see Fig. 14.10). The subduction and mountain building also included volcanic eruptions, the intrusion of granite batholiths and the formation of major faults such as the **Highland Boundary** and **Great Glen** (see Fig. 10.19 on page 99).

Even as the mountains were rising, new sediments were being eroded from them. This material was transported onto the narrow remains of the continental shelf and into the trench above the subduction zone beyond. Deep-water shales and greywacke sandstones from these environments can now be seen as part of the Ordovician sequence in the Southern Uplands of Scotland.

In total, Ordovician events along the Laurentian side of Iapetus may be likened to the more recent subduction of the Nazca Plate beneath South America and the formation of the Andes Mountains.

England/Wales/southern Ireland

Sedimentation here continued in a similar pattern to Cambrian times, with shales in offshore basins and shallow-water deposits nearer to the shores of the Baltic Continent. However, during the Ordovician, widespread volcanic activity added large amounts of lava, tuff and agglomerate to the sequence. These volcanic rocks now stand out as the highest, most resistant, hills of the Lake District, Southern Ireland and Wales (for example, Snowdonia).

As Fig. 14.9 shows, the volcanism is thought to have been caused by subduction but there is some argument about where this actually took place. Possibly the earlier subduction zone (which origi-

Fig. 14.9 Iapetus in the Ordovician period

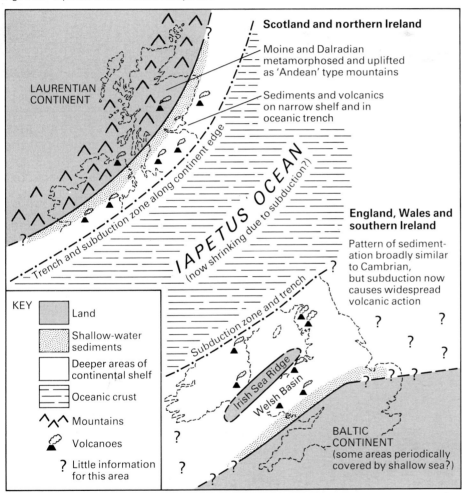

Scotland and northern Ireland
Moine and Dalradian metamorphosed and uplifted as 'Andean' type mountains
Sediments and volcanics on narrow shelf and in oceanic trench

LAURENTIAN CONTINENT

Trench and subduction zone along continent edge

IAPETUS OCEAN (now shrinking due to subduction?)

Subduction zone and trench

England, Wales and southern Ireland
Pattern of sedimentation broadly similar to Cambrian, but subduction now causes widespread volcanic action

Irish Sea Ridge
Welsh Basin

BALTIC CONTINENT (some areas periodically covered by shallow sea?)

KEY
Land
Shallow-water sediments
Deeper areas of continental shelf
Oceanic crust
∧∧ Mountains
🌋 Volcanoes
? Little information for this area

Section across ocean

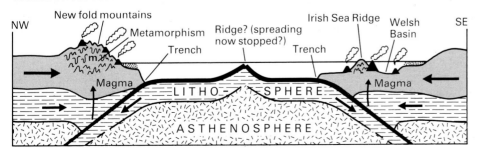

NW
New fold mountains
Metamorphism
Trench
Magma
Ridge? (spreading now stopped?)
Trench
Irish Sea Ridge
Welsh Basin
SE
Magma
LITHO - SPHERE
ASTHENOSPHERE

nally formed the Irish Sea Ridge during the Pre-Cambrian) became active again, or (more likely) a new zone developed to the north of this. Notice that the volcanoes formed islands in the seas of the continental edge and did not rise directly from oceanic crust: for this reason they should not be thought of as an island are of the type described on page 142.

Fig. 14.11 Resistant Ordovician volcanic rocks at the summit of Glyder, Snowdonia

The closure of Iapetus in the Silurian

(435 to 395 Ma: Fig. 14.12)

By the early Silurian, subduction had reduced Iapetus to a narrow ocean, and the rest of the period marks the final stages of closure.

Even before total collision took place, a series of parallel land ridges developed as pressure buckled and uplifted parts of both continental edges. Ridges such as **Solwayland** and **Cockburnland** stood between the mainlands and caused the pattern of deposition to become more complex by each supplying sediments to their adjacent flooded basins. The shoreline of the main Baltic Continent was also changing as a rise in sea level allowed shallow-water sediments to spread across what had previously been land areas in the Midlands and South East England.

By the late Silurian, Iapetus no longer existed. The edges of the Laurentian and Baltic continents (including, of course, the two halves of Britain) had merged to-

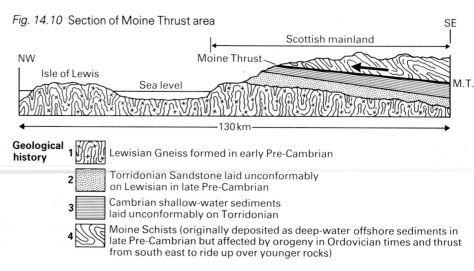

Fig. 14.10 Section of Moine Thrust area

Geological history

1. Lewisian Gneiss formed in early Pre-Cambrian
2. Torridonian Sandstone laid unconformably on Lewisian in late Pre-Cambrian
3. Cambrian shallow-water sediments laid unconformably on Torridonian
4. Moine Schists (originally deposited as deep-water offshore sediments in late Pre-Cambrian but affected by orogeny in Ordovician times and thrust from south east to ride up over younger rocks)

Fig. 14.12 Iapetus in the Silurian period

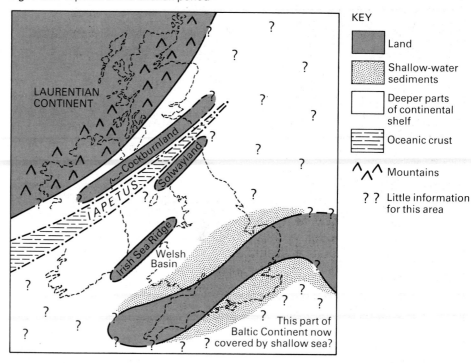

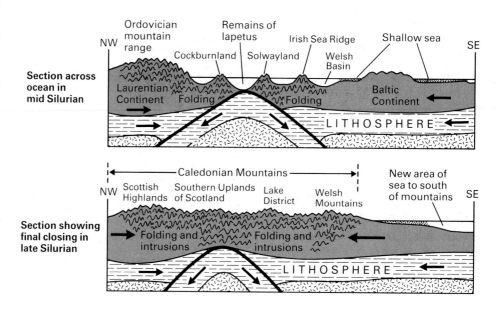

gether and raised the mountains of the **Caledonian Fold Belt**. However, these final phases of collision and orogeny were not as powerful as might have been expected. Lower Palaeozoic rocks of England, Wales, southern Scotland and most of Ireland show only low-grade regional metamorphism (for example, Welsh slate) and although folding is obvious it does not usually involve overfolding or thrusting. Certainly the forces of this Silurian deformation were less intense than those that had occurred earlier when the Scottish Moine and Dalradian groups were affected during the Ordovician. Perhaps the final stages of the orogeny were relatively 'weak' because, even as collision took place, the pattern of continents was changing. A new ocean called the **Rheic** was by then beginning to open to the south of Britain.

The total effect of the Caledonian Orogeny was to produce a range of mountains whose features are still evident today. (Note: by using the word 'total' we are including the complete series of events which lead up to collision as well as the effect of the collision itself.) Fig 14.5 shows how fragments of the Caledonian Fold Belt can be seen in highland areas of Canada, Greenland and Scandinavia. Fig. 14.13 shows the main Caledonian features in Britain. Notice that the trend of the fold axes is north east to south west; this shows that the original pressure of the orogeny must have acted from the north west and south east.

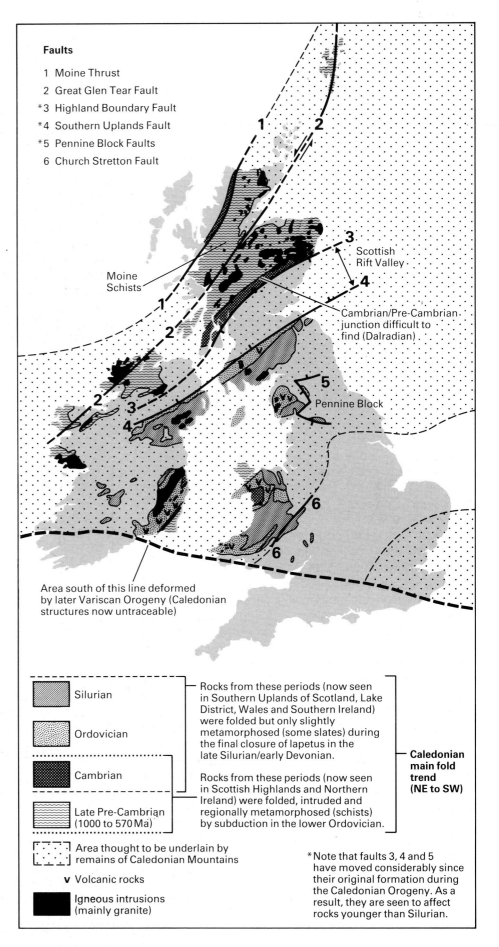

Faults

1 Moine Thrust
2 Great Glen Tear Fault
*3 Highland Boundary Fault
*4 Southern Uplands Fault
*5 Pennine Block Faults
6 Church Stretton Fault

Scottish Rift Valley

Cambrian/Pre-Cambrian junction difficult to find (Dalradian)

Pennine Block

Moine Schists

Area south of this line deformed by later Variscan Orogeny (Caledonian structures now untraceable)

Silurian

Ordovician

Cambrian

Late Pre-Cambrian (1000 to 570 Ma)

Rocks from these periods (now seen in Southern Uplands of Scotland, Lake District, Wales and Southern Ireland) were folded but only slightly metamorphosed (some slates) during the final closure of Iapetus in the late Silurian/early Devonian.

Rocks from these periods (now seen in Scottish Highlands and Northern Ireland) were folded, intruded and regionally metamorphosed (schists) by subduction in the lower Ordovician.

Caledonian main fold trend (NE to SW)

Area thought to be underlain by remains of Caledonian Mountains

v Volcanic rocks

Igneous intrusions (mainly granite)

*Note that faults 3, 4 and 5 have moved considerably since their original formation during the Caledonian Orogeny. As a result, they are seen to affect rocks younger than Silurian.

Fig. 14.13 Distribution of late Pre-Cambrian, Cambrian, Ordovician and Silurian rocks, and the effects of the Caledonian Orogeny

Episode 3: Britain as part of the Caledonian Continent

With the closing of Iapetus, the Baltic and Laurentian regions became a single landmass known as the **Caledonian Continent**. To the south of this, a new ocean called the **Rheic** began developing with part of its shoreline extending across southern Britain.

This episode traces the story through Devonian and Carboniferous times until the eventual closing of the Rheic Ocean caused the **Variscan Orogeny** by bringing Southern Europe into contact with the Caledonian Continent. Throughout these ages climate had an important effect on geology because, by that time, the northward drift had carried Britain into equatorial latitudes.

Devonian events (395 to 345 Ma: see Fig. 14.14)

For this period it is convenient to divide Britain by an east–west line from Bristol to London. This marks the approximate shoreline of the Rheic Ocean.

Northern area (land)

This area was dominated by the newly formed mountains. The last phases of the Caledonian Orogeny were still taking place when the period began but, as time passed, erosion became the most important process. During the Devonian, thousands of metres of Lower Palaeozoic and late Pre-Cambrian rocks were worn away.

Not all the eroded sediment reached the sea. Some of it was deposited on land to produce a series of rocks known as the **Old Red Sandstones**. A large amount of Old Red Sandstone was laid down by rivers flowing into lower-lying areas between the main peaks. Such areas are called **intermontaine basins**. As Fig. 14.14, shows, one of these basins lay in the region of the Orkney and Shetland Islands. Here the sequence includes mudstones which were deposited in a lake (fossils of fresh-water fish have been found). Another intermontaine basin lay between the Highland Boundary and Southern Uplands faults. As sandstones were deposited here, the whole area subsided to produce a **rift valley** stretching from central Scotland to northern Ireland. Numerous volcanic eruptions accompanied the subsidence.

Other deposits of Old Red Sandstone were formed to the south of the mountains, on flood plains and deltas close to the Rheic shore. These sediments, now found in parts of South Wales and southern Ireland, show features such as seasonal deposition and dessication cracks. This evidence supports the idea that Britain was then situated about 15–20° south of the equator and had only limited seasonal rainfall.

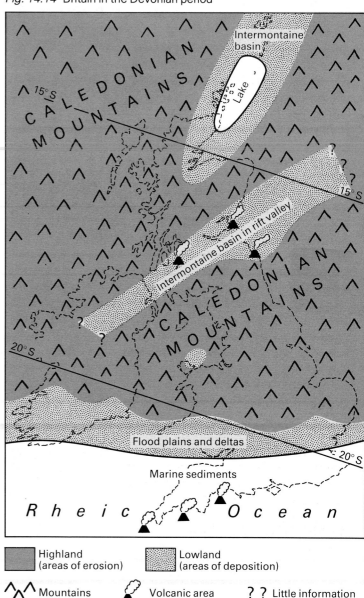

Fig. 14.14 Britain in the Devonian period

Symbol	Description
Highland (areas of erosion)	Lowland (areas of deposition)
Mountains	Volcanic area
? ? Little information	

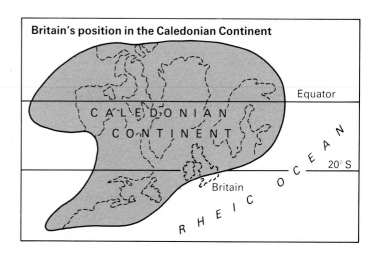

Britain's position in the Caledonian Continent

Fig. 14.15 Devonian sandstone at Duncansby Head near John o'Groats, north east Scotland

Southern area (sea)

This was an area of continental shelf beneath the Rheic Ocean but the exact shoreline must have varied. At times the sea spread northwards to leave marine sediments interbedded with the Old Red Sandstones of South Wales. At other stages it must have retreated south because delta deposits are known from parts of Devon.

The sea floor also seems to have varied in depth; deeper subsiding areas (for example, west Cornwall) received thousands of metres of muds and greywackes, while other regions (for example, north Devon) acquired shallow-water limestones with fossil corals and brachiopods.

Carboniferous events

(345 to 280 Ma: See Fig. 14.17)

Carboniferous rocks and fossils show that northward drift had moved Britain (as part of the Caledonian continent) close to the equator. It is convenient to discuss this period in two parts:
- Lower Carboniferous, a time of warm seas and limestones;
- Upper Carboniferous, a time of deltaic conditions with gritstones and coal deposits.

Lower Carboniferous

Although the main Rheic Ocean still lay across the south of Britain, a rise in sea level meant that its waters now spread slowly north over the remains of the Caledonian Mountains. This produced a major unconformity with Carboniferous sediments being deposited on folded eroded Lower Palaeozoic rocks (see Fig. 10.27 on page 102).

The warm tropical environment of the Lower Carboniferous sea favoured **limestone** deposition but the pattern of sedimentation also depended on the shape of the drowned landscape beneath. Some high areas remained as islands throughout and did not receive sediment (for example, **St George's Land**). The deeper waters of flooded valleys and basins tended to accumulate beds of muddy limestone and shale. Only in the really shallow and warmest regions of the sea was pure grey/white limestone formed.

Lower Carboniferous (shallow-water) limestones can now be seen in the Northern Pennines, Peak District, Mendips and much of central Ireland. They provide an interesting landscape with caves, potholes and underground streams (Fig. 6.25 on page 53) and the rocks are well known for the fossils (for example, brachiopods, crinoids and goniatites) they contain. The presence of colonial corals and algal reefs shows that tropical conditions existed during this time.

Although the limestone sea spread as far as the Scottish Rift Valley, this particular area also received sediment from the remaining Caledonian Mountains to the north. As time passed, the deposition of sands and muds produced a series of deltas along this shore.

Volcanic activity also occurred during the Lower Carboniferous. Basalts were erupted in central Scotland (for example, Arthur's Seat in Edinburgh is the eroded remains of a volcanic cone) and lavas and tuffs have been found in the Scottish Borders and Derbyshire.

Fig. 14.16 Carboniferous limestone landscape, northern Pennines

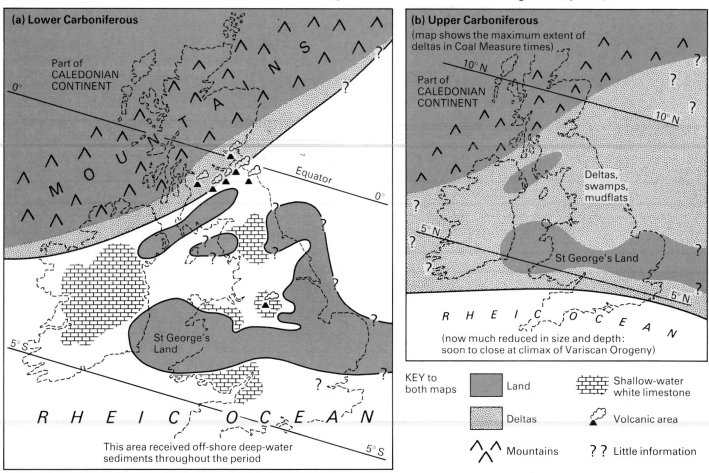

Fig. 14.17 Britain in the Carboniferous period (both maps are very generalised as sea level varied throughout the period)

(a) Lower Carboniferous

Part of CALEDONIAN CONTINENT

0°

Equator

0°

St George's Land

5° S

R H E I C   O C E A N

This area received off-shore deep-water sediments throughout the period

5° S

(b) Upper Carboniferous
(map shows the maximum extent of deltas in Coal Measure times)

10° N

Part of CALEDONIAN CONTINENT

10° N

Deltas, swamps, mudflats

5° N

St George's Land

5° N

R H E I C   O C E A N

(now much reduced in size and depth: soon to close at climax of Variscan Orogeny)

KEY to both maps

Land

Deltas

Mountains

Shallow-water white limestone

Volcanic area

?? Little information

Upper Carboniferous

Deep-water sediments continued to be deposited in the Rheic Ocean across southern England but, over the rest of the country, conditions were changing. A series of uplifts (presumably early effects of the Variscan Orogeny) reduced the depth of water lying across central Britain and Ireland and increased the supply of sediment to the northern deltas. As a result, these deltas began to infill the shallow water and extend south towards St George's Land. The advancing deltas produced an interbedded sequence of shales and coarse arkose sandstones known as **Millstone Grit**. These rocks now give a distinctive landscape to the dark moorlands of the mid Pennines (Fig. 14.18).

Mid-way through the Upper Carboniferous, the deltas had become so extensive that only the Rheic Ocean (across southern England) remained a truly marine area. Everywhere else (except higher regions such as St George's Land and the Caledonian Mountains of northern Scotland: see Fig 14.17b) had become a vast swampy deltaic area of river channels, lakes and coastal mudflats. It was this environment which provided Britain's most valuable geological deposit: the **Coal Measures**.

Newly evolved species of land plants grew rapidly in the hot wet equatorial climate, and the swampy conditions allowed vegetation to sink and accumulate rather than rot away. However, beds of coal (**seams**) would not have been formed unless, every so often, changes in the pattern of deposition caused other sediments to bury and preserve the layers of plant material.

Fig. 14.18 Millstone grit landscape near Oldham, Lancashire

To work out how the patterns of deposition were changing we must study sequences of rock which contain coal seams. Although there are many slight variations, Coal Measure sediments usually occur in the order shown by Fig. 14.19. Look carefully at this diagram and note how each complete sequence of rocks (called a **cyclothem**) can be explained in terms of the sediments produced by a delta building out across a low-lying lake or lagoon. Notice also that the same sequence of rocks is repeated again and again throughout the Coal Measures; this suggests that, when deltas had developed to a stage where forests were growing on their surface, they were eventually flooded over and the whole process of building out had to begin again. Possible explanations for the periodic 'drowning' of deltas include the following.

Subsidence. Modern deltas tend to subside slowly and only remain above water if enough sediment is being supplied to keep building them up. If the supply of sediment to part of the delta is reduced (for example, by a storm altering the pattern of distributaries), then that area is likely to be 'drowned' and become a lake or lagoon.

Rise of sea level. In some cyclothems, marine fossils such as brachiopods or fish have been found (rather than the more usual fresh-water bivalves) in the shales immediately above the coal seam. This evidence shows that some deltas were 'drowned' by rising sea level. Fluctuations of sea level during the Upper Carboniferous could have been caused by alternate periods of freezing and thawing in the ice sheets which covered much of **Gondwanaland** at that time (see page 147).

Fig. 14.19 Generalised sequence of Coal Measures sediments

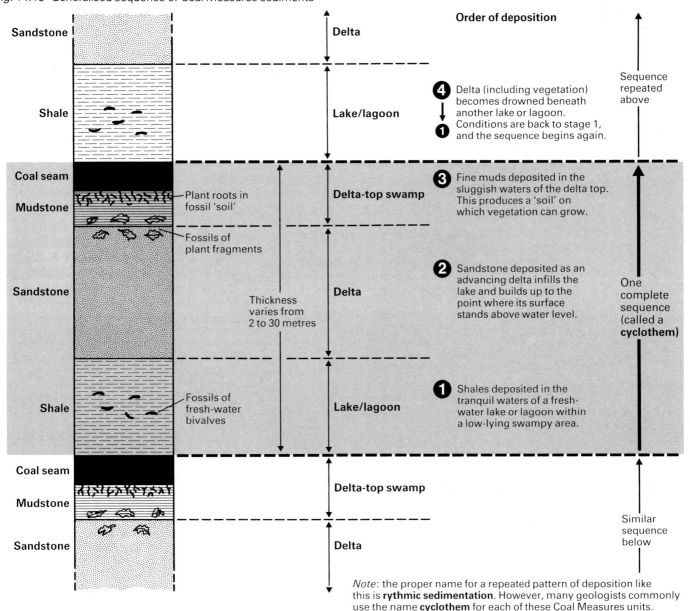

Order of deposition

Sequence repeated above

4 Delta (including vegetation) becomes drowned beneath another lake or lagoon. Conditions are back to stage 1, and the sequence begins again.

3 Fine muds deposited in the sluggish waters of the delta top. This produces a 'soil' on which vegetation can grow.

2 Sandstone deposited as an advancing delta infills the lake and builds up to the point where its surface stands above water level.

1 Shales deposited in the tranquil waters of a fresh-water lake or lagoon within a low-lying swampy area.

One complete sequence (called a **cyclothem**)

Similar sequence below

Sandstone — Delta

Shale — Lake/lagoon

Coal seam

Mudstone — Delta-top swamp — Plant roots in fossil 'soil'

— Fossils of plant fragments

Sandstone — Delta — Thickness varies from 2 to 30 metres

Shale — Lake/lagoon — Fossils of fresh-water bivalves

Coal seam

Mudstone — Delta-top swamp

Sandstone — Delta

Note: the proper name for a repeated pattern of deposition like this is **rythmic sedimentation**. However, many geologists commonly use the name **cyclothem** for each of these Coal Measures units.

The Variscan Orogeny: late Carboniferous/early Permian (290 to 280 Ma)

This orogeny was caused as closure of the Rheic Ocean brought Southern Europe into contact with the Caledonian Continent. The most powerful deformation occurred in central France and southern Germany but its effects can also be clearly seen in south west Britain. (Note: partly because it has been studied and written about in several European countries, the Variscan has two alternative names: the **Armorican** or **Hercynian**.)

As Fig. 14.20 shows, Britain's main Variscan folds are in Devon, Cornwall and the most southern parts of Wales and Ireland. Folds in these areas tend to be tight, vertical or even overturned (see Fig. 10.10 on page 96) and have their axes running in an east-west direction. This evidence shows that intense forces must have been acting north-south as the orogeny compressed Devonian and Carboniferous rocks against the edge of the Caledonian Continent. Regional metamorphism, the **Lizard Thrust** fault and the south west granite batholith provide further evidence of Variscan mountain building in this area.

North of the main fold belt the deformation was less intense and the east-west fold trend less obvious. Here, Variscan forces tended to cause new movements on older buried Caledonian structures which in turn produced folding in the overlying sediments. For example, movement on the **Pennine Block Faults** uplifted the centre of the area to produce a broad anticline structure and renewed subsidence of the **Scottish Rift Valley** formed a syncline.

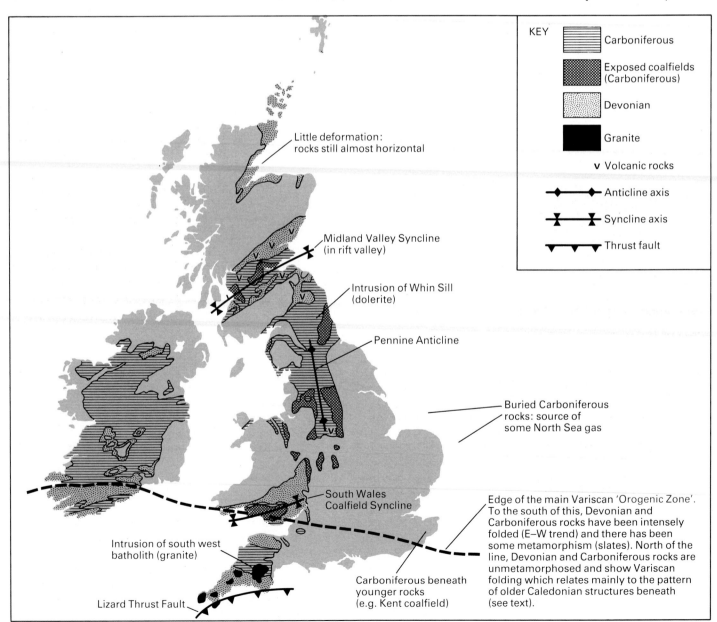

KEY
- Carboniferous
- Exposed coalfields (Carboniferous)
- Devonian
- Granite
- v Volcanic rocks
- Anticline axis
- Syncline axis
- Thrust fault

Little deformation: rocks still almost horizontal

Midland Valley Syncline (in rift valley)

Intrusion of Whin Sill (dolerite)

Pennine Anticline

Buried Carboniferous rocks: source of some North Sea gas

South Wales Coalfield Syncline

Intrusion of south west batholith (granite)

Lizard Thrust Fault

Carboniferous beneath younger rocks (e.g. Kent coalfield)

Edge of the main Variscan 'Orogenic Zone'. To the south of this, Devonian and Carboniferous rocks have been intensely folded (E–W trend) and there has been some metamorphism (slates). North of the line, Devonian and Carboniferous rocks are unmetamorphosed and show Variscan folding which relates mainly to the pattern of older Caledonian structures beneath (see text).

Fig. 14.20 Distribution of Devonian and Carboniferous rocks, and the effects of the Variscan Orogeny

Episode 4: Britain as part of a large landmass with fluctuating sea level

The closure of the Rheic was part of a series of plate movements which (by early in the Permian period) had brought all the world's continents together as the huge landmass called **Pangaea** (see page 147). Although Pangaea became separated into smaller continents during the Mesozoic era, it was not until Tertiary times (when the North Atlantic began opening) that Britain was once again placed near a proper ocean or an active plate margin.

From the Permian to the Tertiary Britain's geological development was greatly influenced by its position within a large continental unit. Although seas did periodically flood the area, they were continental seas rather than real spreading and subduction oceans like the earlier Iapetus or Rheic. (Note: a continental sea is one which covers only continental crust, like, for example, the present-day North Sea.)

Fig. 14.21 Britain in the Permian and Triassic periods

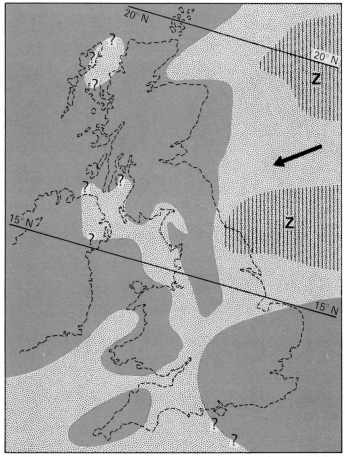

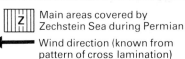

Highland (areas of erosion)	Main areas covered by Zechstein Sea during Permian
Lowland (deposition in desert basins)	Wind direction (known from pattern of cross lamination)
	? ? Little information

See Fig. 14.31 (page 169) for a map showing the distribution of Britain's Permian, Triassic, Jurassic, Cretaceous and Tertiary rocks.

Permian and Triassic events

(280 to 195 Ma: see Fig. 14.21)

During these periods Britain was about 15–20° north of the equator and far from an open ocean (a position roughly similar to that of present-day North Africa/ Arabia). The climate was arid and the most common rocks to form were red desert sandstones. Since these rocks contain very few fossils, correlating the Permian and Triassic is rather difficult. Many geologists prefer to deal with both periods as one overall unit called either the **Permo–Trias** or **New Red Sandstone**. (Note how this second name distinguishes these rocks from the somewhat similar Old Red Sandstones of earlier Devonian age.)

Permian

As the period began, lowlands and basins in and around the newly formed Variscan Mountains became deserts of wind-blown sand. The largest area of deposition extended around the flanks of the Pennines to include the Midlands and much of what is now the North Sea. Smaller basins existed in parts of Devon, Somerset and Cumbria. Red **cross-laminated** (dune) sandstones, like those shown in Fig. 7.26 (page 75), were the main deposit but some breccias and conglomerates were produced by alluvial fans along the steeper slopes of basin edges.

Later in the period, part of the desert area was flooded by the **Zechstein Sea**. This shallow body of water extended from Britain to East Germany and, in the arid conditions of the time, high rates of evaporation produced beds of **gypsum** and **halite**. A modern example of a 'Zechstein-type' environment would be the salt flats around the Persian Gulf (see notes on **sabkhas** on page 73).

Triassic

By the start of this period the Zechstein Sea had retreated (or been completely evaporated) and red wind-blown sandstones were forming in the desert basins once more. As time passed, a continental sea spread across much of Europe. Although this failed to reach Britain, its presence caused more rainfall and in the Mid Triassic a variety of water-lain sediments were deposited. These include pebbly sandstones (laid down by rivers) and red mudstones (produced in shallow lakes). Although conditions remained generally rather dry, the extra rainfall allowed more species of plants and animals (for example, early dinosaurs) to live in the semi-desert environment of this time.

Towards the end of the Triassic, high rates of evaporation returned and low-lying areas such as

Cheshire and north east Yorkshire became sabkha environments were halite was precipitated.

Although a large amount of sediment was deposited in the lowlands during the Permian and Triassic, there was also a great deal of erosion during these periods. Both the Variscan Mountains of southern Britain and the remaining older Caledonian Mountains of the north were greatly reduced in height.

Jurassic events (195 to 135 Ma : Fig. 14.22)

Conditions changed considerably with the start of this period. Britain had now moved north from desert latitudes to lie about 35° north of the equator (the latitude of the present Mediterranean), and another continental sea was beginning to advance across Europe.

Fig. 14.22 Britain in the Mid Jurassic. Sea level changes meant that the coastline varied considerably during this period: see text for details

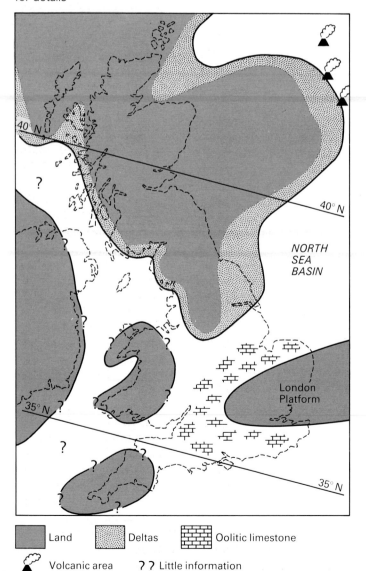

Land   Deltas   Oolitic limestone

Volcanic area   ?? Little information

By the early Jurassic this sea had reached southern Britain and, as the waters deepened and spread north, fine muds were deposited. This sediment produced the **Lias Clays** which are famous for their fossil ammonites, belemnites and marine reptiles such as *Ichthyosaurs* and *Plesiosaurs*. Good exposures of these rocks can now be seen in the soft unstable cliffs of Whitby (Yorkshire) and Lyme Regis (Dorset: see Fig. 14.23). **Sedimentary ironstones**, such as those now found in Lincolnshire and Northamptonshire, were deposited in some shallower parts of the Lower Jurassic sea and a large island called the **London Platform** occupied much of south east England at this time.

As sea level dropped during the Mid Jurassic the waters across southern and central England became shallow and warm. These conditions produced the **oolitic limestones** (page 70 and Fig. 7.15) which can now be seen, for example, in the Cotswolds. To the north, deltas spread into the shallow sea from the remaining areas of highland. These deltaic sediments are now found in North Yorkshire and parts of north west and north east Scotland. They contain sandstones, shales and the occasional thin seam of coal.

Fig. 14.23 Unstable cliffs of Lias Clay near Lyme Regis, Dorset

By the start of the Upper Jurassic, sea level was rising again. This time the sea reached its maximum extent for the period and deposited the **Oxford Clays** and the **Kimmeridge Clays**. Both have proved economically important: Oxford Clays are used for brick making (see page 69) and Kimmeridge Clays are thought to be the **source rock** from which much North Sea oil has come (see page 182).

Towards the end of the Jurassic another lowering of sea level took place. This allowed the London Platform to link with the northern highlands and create a continuous barrier of land. On one side of this was a sea where sandstones and mudstones were forming (this area was approximately in the position of the present North Sea). On the other side of the land barrier a warmer shallower sea left limestones (for example, the **Portland Limestone**: Fig. 14.24) across southern England.

The interpretation of Britain's Jurassic geology has been greatly helped by the presence of ammonite zone fossils. (An example of Jurassic correlation is given in Fig. 12.8 on page 118).

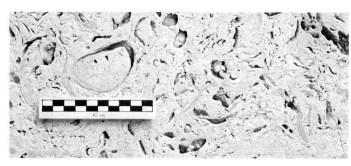

Fig. 14.24 Close-up of Portland Limestone

Cretaceous events (135 to 65 Ma: Fig. 14.25)

The period began (as the Jurassic had ended) with two main areas of deposition. Sands and clays continued in the North Sea Basin but, by now, a freshwater lake had replaced the limestone sea south of the land barrier. This was the **Wealden Lake** and around it a mixed sequence of river sands and muds formed in a swampy environment. These sediments contain types of plant and dinosaur fossils which suggest that Britain still had a very warm climate (latitude was then approximately 40° N).

Near the end of the Lower Cretaceous the land barrier was flooded over when sea level rose once more. At first, the Wealden Lake was covered by beds of marine sandstone but, as the water deepened and spread, the **Gault Clay** was deposited across south east England.

By the start of the Upper Cretaceous a world-wide rise in sea level meant that nearly all of Britain (except possibly the highest Scottish mountain tops) became covered by the **Chalk Sea**. Chalk is a rather unusual limestone made almost entirely of coccoliths (minute calcite discs from the skeletons of algae: see page 70). At the present time, such deposits only form on abyssal plains. However, it is *not* thought that Britain had become a deep oceanic floor during the Upper Cretaceous. The area was still a continental sea but, because so little land was left above water, hardly any normal sediment (sand, mud, etc.) could be eroded or supplied. Practically the only material settling on the sea bed was coccoliths from algae which flourished in such warm conditions. By a slow but continuous

Fig. 14.25 Britain in the Cretaceous Period

(a) Lower Cretaceous:

Relatively low sea level meant that several areas of deposition were separated by land.

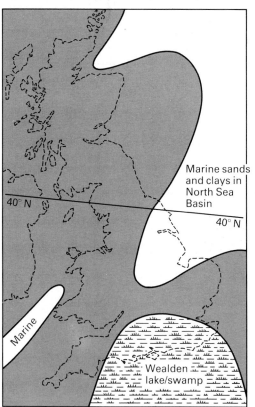

Marine sands and clays in North Sea Basin

Wealden lake/swamp

Marine

40° N

(b) Upper Cretaceous:

Very high sea level meant that only the highest parts of Scotland stood above the Chalk Sea.

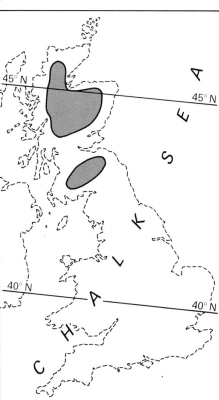

CHALK SEA

45° N

40° N

Fig. 14.26 Chalk downs, Westbury, Wiltshire

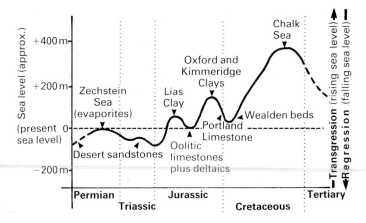

Fig. 14.27 Permian to Cretaceous sea level changes

process they eventually produced up to 1000 m of chalk. About 30 years was needed to give each millimetre.

Chalk is now exposed in the Chilterns, the North and South Downs and the 'white cliffs' of south east England. It is quarried for cement making and also provides supplies of underground water (see page 186). The rock is well known for fossil echinoids and bivalves, plus the nodules and layers of flint it contains. Some flint probably formed from the remains of sea-floor sponges which built their skeletons of silica.

Sea level changes

As you will have realised, Britain's geological development from the Permian to the Cretaceous was largely controlled by the many changes of sea level that took place. Fig. 14.27 summarises these changes in a graph and reminds you of the main rocks that were formed. A possible explanation of events may be found in the plate movements which were occurring at the time.

Page 148 suggested that world sea level is affected by the number of spreading oceanic ridges and the rate at which they are developing. Taking this idea into account, it seems probable that Mesozoic changes in sea level could have been caused by stages in the growth of new oceans that were then separating Pangaea into smaller continents. Of course, the overall pattern may have been complicated by other factors. For example, some continental areas may have subsided and allowed the sea to flood over them without any change in its own level at all.

Tertiary events

(65 to 2 Ma: Fig. 14.28)

This period began with yet another major lowering of sea level, and most of Britain emerged as land once more. However, the North Sea Basin remained below the waves, and here up to 3 km of sediments were deposited during the period. Subsidence and burial of the older rocks (some of which were rich in organic remains) beneath this material is thought to have caused the formation of some **North Sea oil** (see page 182). The exact coastline of the North Sea Basin varied during the Tertiary and at times marine conditions spread into south east England to deposit rocks such as the **London Clay**.

Fig. 14.28 Britain in the Tertiary period

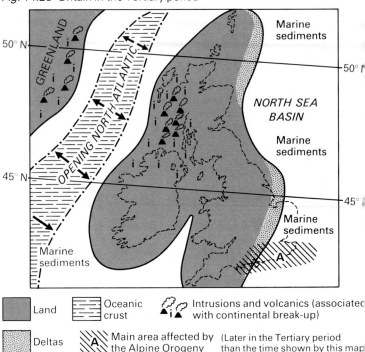

While sedimentary rocks were being deposited in the North Sea, the other side of Britain was affected by a constructive plate margin. As part of the break-up of Pangaea, the Atlantic Ocean had begun opening between Southern Europe and America during the Jurassic. By early in the Tertiary (60 Ma), the spreading ridge had developed far enough north to begin splitting Northern Europe from Greenland. As this took place, magmas from the ridge rose into the separating continental edges. Similar igneous rocks of Tertiary age can now be seen on either side of the North Atlantic. The basalt flows of Northern Ireland (for example, Giant's Causeway) and the volcanic rocks and intrusions of Scottish islands such as Skye and Mull are of the same general age and type as the igneous rocks of east Greenland.

The opening of the North Atlantic ended a 350 million year old geological link between Europe and North America. However, the new break was not exactly along the line of the Caledonian collision which had originally brought them together (as the Baltic and Laurentian continents) in late Silurian times. In effect, Scotland and Northern Ireland were 'left behind' on the European side.

Another series of Tertiary plate movements left its impression on south east England when, between 25 and 30 Ma, the **Alpine Orogeny** took place. Britain was very much on the fringe of this event so major features such as regional metamorphism or large granite batholiths did not occur here. However, folds such as the **Wealden Anticline** were produced, and many of the older fault lines moved again in response to the pressure. As its name suggests, the Alpine Orogeny was centred in Southern Europe; it was brought about by the northward movement of Africa and occurred at about the same time as the Himalayan Orogeny brought India into contact with Asia.

Fig. 14.29 McLeod's Tables, Isle of Skye, Scotland, Flat-topped mountains produced by erosion of horizontal basalt lava flows (Tertiary age)

Fig. 14.30 Alpine folding at Stair Hole, Dorset ▶

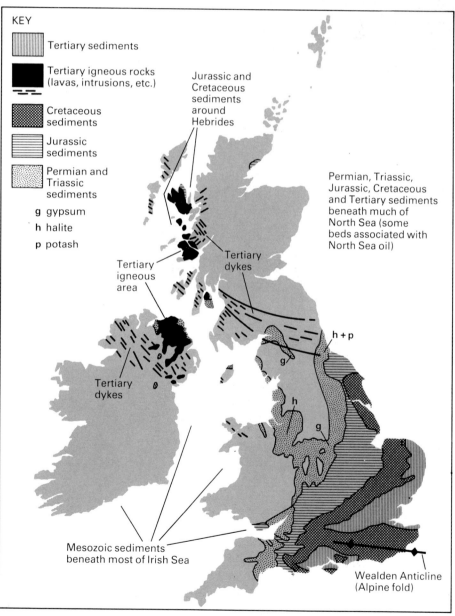

KEY

Tertiary sediments

Tertiary igneous rocks (lavas, intrusions, etc.)

Cretaceous sediments

Jurassic sediments

Permian and Triassic sediments

g gypsum
h halite
p potash

Jurassic and Cretaceous sediments around Hebrides

Permian, Triassic, Jurassic, Cretaceous and Tertiary sediments beneath much of North Sea (some beds associated with North Sea oil)

Tertiary igneous area

Tertiary dykes

Tertiary dykes

Mesozoic sediments beneath most of Irish Sea

Wealden Anticline (Alpine fold)

Fig. 14.31 Distribution of Permian, Triassic, Jurassic, Cretaceous and Tertiary rocks

Episode 5: the final shaping of the British Isles

The opening of the North Atlantic and the Alpine Orogeny were the last major structural events to affect Britain. By Mid Tertiary times the basic pattern of solid (bedrock) geology had been firmly established. Since then the 'final shaping' has taken several forms.

Tertiary tilting and erosion

The combined effect of continental rifting on one side and a subsiding North Sea Basin on the other produced a general tilting of Britain towards the south east. The rising western and northern areas suffered a great deal of erosion during the Upper Tertiary and their covering of Mesozoic sediments, such as the widespread Cretaceous Chalk, was removed to re-expose older Palaeozoic rocks below.

Climatic change and the Ice Age

By the end of the Tertiary, Britain had moved north to about its present latitude. At this time a general lowering of world temperatures took place and the Ice Age began. It was a not single simple event; during the two million years of Quaternary time, ice sheets repeatedly advanced over large areas of North America and Northern Europe. Between glaciations, the climate improved, ice melted and **interglacial** conditions occurred.

Britain was affected by at least four periods of glaciation but it is difficult to gain much information about any of them except the most recent. Each time new ice sheets advanced they destroyed or modified the effects of any previous glacial action. Fig. 14.32 shows the extent of the last glacial advance but to picture the situation properly you must realise that the ice was up to 1 kilometre thick.

The Ice Age, together with its associated changes in sea level and movements of meltwater, has shaped much of our present landscape. In particular, highland areas were eroded and the material re-deposited across much of the lowlands. (See pages 59 to 61 where the geological work of ice was discussed in more detail.)

In some senses the effects of glaciation are not yet over. For example, northern Scotland is still rising by isostatic adjustment as it recovers from the weight of ice (page 17). It should also be mentioned that our present climate probably represents another interglacial period. Given enough time (it is now 10 000 years since the last British ice sheets melted) will glaciers advance once more?

A look into the future: episode 6?

Apart from wondering whether the ice sheets are about to return, let us consider something rather more long term. No ocean can go on spreading for ever, so eventually the Atlantic floor must begin to subduct back into the asthenosphere. Is it possible that, millions of years from now, a subduction zone will develop somewhere off the edge of Western Europe? Only time will tell!

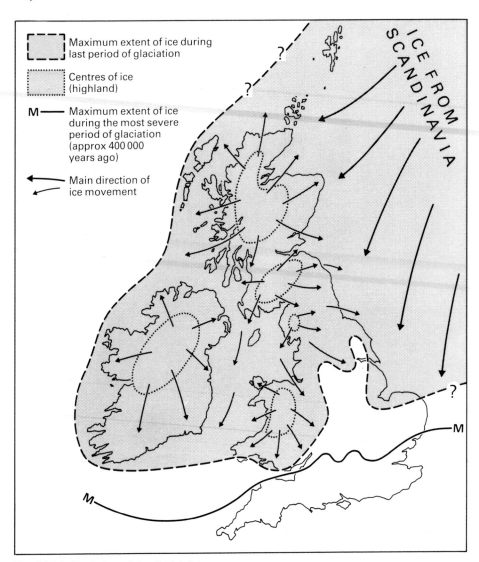

Maximum extent of ice during last period of glaciation

Centres of ice (highland)

M — Maximum extent of ice during the most severe period of glaciation (approx 400 000 years ago)

Main direction of ice movement

ICE FROM SCANDINAVIA

Fig. 14.32 Glaciation of the British Isles

Questions

For questions 1–8 write the letter (A,B,C, . . .) of the correct answers in your notebook.

1 Which **one** of the following has caused Britain to move to its present latitude?
A isostasy   B changing climates
C the cycle of rocks
D spreading and subduction of plate material
E convention currents in the oceans

2 Which **two** of the following describe Britain's geological situation/position during the Lower Palaeozoic era?
A separated into two parts by the Rheic Ocean
B situated in the southern hemisphere
C separated into two parts by the Atlantic Ocean
D situated in the northern hemisphere
E separated into two parts by the Iapetus Ocean

3 In Britain the Variscan Orogeny mainly affected which **one** of the following?
A Mesozoic rocks in Ireland
B Tertiary rocks in Wales
C Jurassic rocks in Yorkshire
D Pre-Cambrian rocks in Scotland
E Upper Palaeozoic rocks in south west England

4 Which **one** of the following describes Britain's position during the Carboniferous period?
A about 40° north of the equator
B about 50° south of the equator
C close to the equator
D close to the Arctic Circle
E in the middle of the Iapetus Ocean

5 The oldest rock outcropping in Britain is
A a Pre-Cambrian dyke   B Silurian slate
C Devonian granite   D Pre-Cambrian gneiss
E Cambrian sandstone

6 Which **two** of the following statements about the Ordovician are false?
A Graptolites provide useful zone fossils.
B Rocks of this age normally show Alpine folding.
C Volcanic rocks of this age are not found in Britain
D Rocks of this age can be seen in the Southern Uplands of Scotland.
E The period is part of the Lower Palaeozoic era.

7 During which **three** of the following periods did parts of Britain have a dry arid climate?
A Devonian   B Cretaceous   C Triassic
D Permian   E Silurian

8 During which **one** of the following times was practically all of Britain covered by the sea?
A Lower Cretaceous   B Upper Carboniferous
C Lower Devonian   D Upper Cretaceous
E Middle Tertiary

9 Each of the following statements refers to one of the 12 geological periods (Pre-Cambrian to Quaternary). Copy and complete them.
(a) Oolitic limestones and ironstones formed in the . . . ?
(b) The Ice Age shaped much of Britain's scenery during the . . . ?
(c) Trilobites are the most useful zone fossils for the . . . ?
(d) The Coal Measures were deposited during the . . . ?
(e) The Zechstein Sea deposited evaporites in the . . . ?
(f) The Laxfordian and Scourian orogenies happened during the . . . ?
(g) Old Red Sandstone was deposited during the . . . ?
(h) Iapetus became a narrow ocean then finally closed in the . . . ?
(i) Although a sea spread over parts of Europe Britain remained a desert throughout the . . . ?
(j) The London Clay was formed during the . . . ?
(k) Subduction and compression caused the Moine Thrust during the . . . ?
(l) Billions of coccoliths produced thick beds of pure white limestone during the . . . ?

10 Match the areas in list 1 of the table below with the rock types you would expect to find there (list 2). Make a copy of the table in your notebook with list 2 correctly arranged.

List 1	List 2
Giant's Causeway (N. Ireland)	Jurassic shales with ammonites
North and South Downs	Moine Schists
Peak District (Derbyshire)	Devonian sandstones and mudstones
Orkney Islands	Cretaceous chalk with flint nodules
Scottish Highlands	Carboniferous limestone
Coast of Dorset and North Yorks.	Tertiary basalt with columnar jointing.

11 The following events were all important during Britain's geological history. Arrange them into the correct order (beginning with the oldest).
(a) The Alpine Orogeny takes place.
(b) Britain moves from the southern into the northern hemisphere.
(c) Pre-Cambrian gneisses are formed.
(d) The Caledonian Orogeny causes folding and metamorphism.
(e) Temperature decreases and the Ice Age begins.
(f) Britain moves to the latitude of the present-day Mediterranean Sea.
(g) Britain is part of the supercontinent of Pangaea.
(h) The Iapetus Ocean opens and then begins to close.

(i) Britain is situated close to the shore of the Rheic Ocean.

(j) The Variscan Orogeny takes place.

(k) The North Atlantic Ocean begins to open.

12 Study the map of the British Isles below.

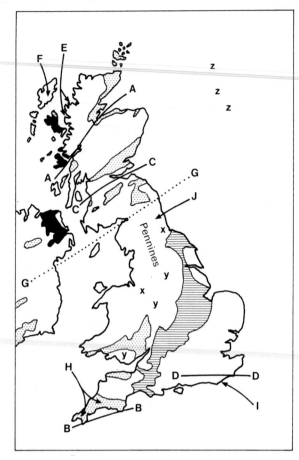

(a) What age of rock is found in the areas shaded with horizontal lines?

(b) What age of rock is found in the areas shaded with dots?

(c) What age are the volcanic rocks of Scotland and Ireland (shaded black on the map)? What was happening in this region at that time to cause such widespread volcanism?

(d) Name faults A–A and B–B; state what type of movement has occurred in each case.

(e) Name folds C–C and D–D; state which orogenies produced them.

(f) What type of erosion was responsible for the lochs in area E; when did this erosion take place?

(g) What age and type of rocks are found on island F; what is it about these rocks that makes them so difficult to study?

(h) What is the significance of dotted line G in terms of Britain's geological development during the Lower Palaeozoic era?

(i) What type of igneous intrusion underlies much of area H? During which orogeny was it formed?

(j) What rock type can be seen in the white cliffs along the coast at I? When was it formed and what were conditions like at the time?

(k) Name the intrusion which underlies much of area J; state what rock type it contains.

(l) Name the mineral extracted from the areas labelled x; at what latitude was Britain situated when these deposits were formed?

(m) What fuel is extracted from the areas labelled y and what fuel is extracted from the areas labelled z?

(n) What age of rock forms the Pennine Hills?

13 The map below shows Britain during part of the Lower Palaeozoic era.

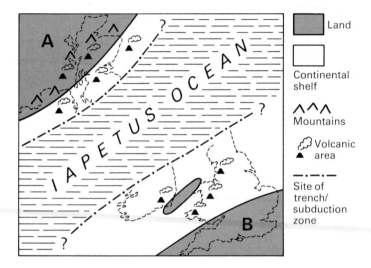

(a) Name continents A and B.

(b) What type of crust existed in the area shaded with dashed horizontal lines?

(c) Note the volcanic areas shown on the map; carefully explain what plate movements are thought to be responsible for these eruptions (draw a section across Iapetus to illustrate your answer).

(d) Describe what happened to the rocks that had been accumulating on the continental shelves when Iapetus finally closed.

(e) Why has it been necessary to place question marks on some parts of the map?

(f) Explain how the distribution of fossil trilobite species has given further evidence about the existence of Iapetus.

14 Obtain a blank map of the British Isles (trace one if necessary). Use this to illustrate the extent of land, sea and other relevant features during any named period chosen from the Upper Palaeozoic or Mesozoic eras. Write an account of the geology of this period.

15 With reference to an area which you have studied in the field explain how a study of rocks, fossils and structures yields information about Britain's geological past.

15 Economic geology

This is the part of our subject concerned with finding, extracting and using geological raw materials. Industries such as metal working, energy production, chemical processing and the manufacture of building materials all depend on a steady supply of reasonably priced minerals, rocks and fuels.

As you would expect from the name, economic geology is also about making money; some of the world's largest and most powerful companies have gained their wealth from exploiting the earth's resources. Politics are involved too, since the economies of many nations would collapse without income from the export of minerals or oil.

Strictly speaking, an economic deposit is one which can be extracted and sold at a profit, but many things affect the profitability of mines and quarries. Geological factors might include how deeply the deposit is buried or how much waste rock has to be taken out. Other, non-geological, considerations could be transport costs or the availability of government subsidies for certain projects. The situation becomes even more complicated when environmental issues are taken into account. There are, for example, some deposits which could be profitable but companies have not been allowed to extract because doing so would ruin a landscape or disrupt nearby settlements.

Apart from the extraction of raw materials, geologists are also employed in economic activities such as finding water supplies and surveying suitable sites for bridges, tunnels and dams.

Finding economic deposits

The best way to find economic deposits is to search for geological situations similar to those where deposits have already been found. For example, when looking for halite or gypsum, a rock sequence of desert sediments is well worth investigating. On the other hand, no-one would spend money drilling through Pre-Cambrian gneiss in search of coal.

Fig. 15.1

AN INTEREST IN ECONOMIC GEOLOGY DOES NOT ALLOW YOU TO SELL YOUR PLACE ON THE FIELD TRIP TO THE HIGHEST BIDDER!!

Before beginning the expensive work of field mapping or drilling in a particular spot, most companies use quicker/cheaper methods to survey a whole region. These methods are designed to discover any 'likely spots' that could be worth investigating more closely. Regional surveys can be based on various techniques, for example

- **seismic techniques**: by sending a series of shock waves (produced by small explosive charges) through rocks it is possible to collect information about the structures and different densities of materials present below the surface (see pages 11 and 12);
- **geochemical techniques**: by collecting and analysing small samples of soils, stream water, etc., it is possible to discover if any areas have a concentration of particular elements;
- **magnetic techniques**: any area with unusually strong magnetism is likely to be rich in iron.

If regional surveys show promising sites, **boreholes** may be drilled to investigate these places more thoroughly. The drills are designed to remove an undamaged central core of rock from each hole so samples can be brought to the surface for study. A series of boreholes is useful not just to prove if particular deposits are actually present, but also to show their extent and indicate any problems that might affect them being extracted.

Look at the borehole data shown in Fig. 15.2. The three cores come from a survey which has found coal seams in rocks of Upper Carboniferous age. Although cores A and B show that the seams are thick enough to mine and are at a reasonably shallow depth, you will notice that the dip of the seams does not match directly across from one site to the other. This fact shows that the rocks must be faulted or folded somewhere between A and B, and suggests that mining the coal could be quite difficult. Core C shows that the seams do not extend very far because no Coal Measures occur at that site. Information like this is vital because it allows companies to calculate the extent and value of underground deposits and estimate the costs of extraction. If the value of a deposit is greater than the cost of getting it out then the deposit is **economic** and the company has a money-making opportunity. Even if a particular deposit does not seem economic, the survey should not be considered a total waste of time; at least the company knows what is there and can keep it in mind for possible extraction in the future should the price of that material ever become high enough to make the operation pay.

Fig. 15.2 Borehole data (see page 173 for details)
(a) *What is the vertical distance between the two coal seams?*
(b) *At what distance below the surface is the Triassic–Silurian junction found in borehole C?*

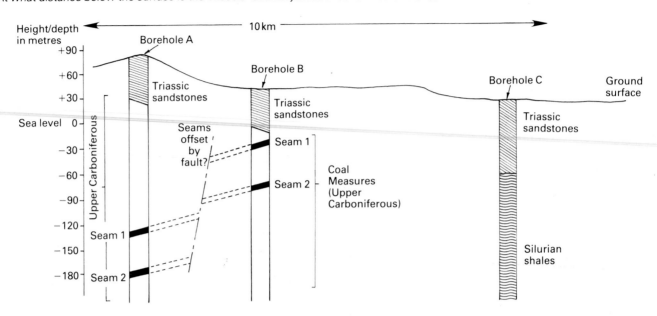

Table 15.1(a) Metals and their uses

Metal	Main ore(s)	Uses
Aluminium (Al)	Bauxite	Strong but lightweight metal resistant to corrosion. Used, for example, in aircraft and building industries. Bauxite can also yield aluminium oxide: a hard substance used in abrasives
Chromium (Cr)	Chromite	Alloyed (mixed) with steel to produce rust resisting stainless steel. Chromium chemicals used in dyes and pigments.
Copper (Cu)	Chalcopyrite Malachite Azurite	Excellent conductor of electricity, widely used in cables, motors, switches, etc. Also good conductor of heat and therefore used for heating pipes, vehicle radiators, etc. Copper is alloyed with zinc to make brass. Copper chemicals are in some dyes and insecticides.
Gold (Au)	Mainly found as pure gold metal	Stored as bullion bars to back up the value of paper money: dug from one hole in the ground (mine) and placed in another (bank vault). Also used for jewellery.
Iron (Fe)	Haematite Magnetite Limonite	A wide variety of different steels are produced by alloying iron with other elements and/or applying special treatments e.g. tungsten steels for cutting tools and low carbon steels for flexible wires and pipes. Iron and steel are obviously the most important metals for industry.
Lead (Pb)	Galena	Soft, dense, easily shaped metal. Used particularly in battery making. Lead chemicals are added to petrol and used in some glass, glazes and paints.
Nickel (Ni)	Pentlandite (nickel-iron sulphide)	Adds strength and hardness to some steel alloys. 'Silver' coins are actually an alloy of nickel and copper.
Silver (Ag)	May occur as pure silver	The majority of silver is obtained as a trace element in lead and copper ores. Used for jewellery and as bullion. Silver chemicals are important in the making of photographic films.
Tin (Sn)	Cassiterite	Used to plate (cover) the inside of steel food cans. Also used in alloys such as bronze (tin and copper) and solder (tin and lead).
Uranium (U)	Pitchblende	Nuclear fuel for power stations and weapons.
Zinc (Zn)	Sphalerite	Widely used for galvanising (a process which gives steel a rust resisting surface by dipping it in molten zinc). Zinc is alloyed with copper to make brass. White, powdered zinc oxide is used in some paints and medicines.

Table 15.1(b) Other economic rocks/minerals and their uses

Amphibole	Asbestos is a form of amphibole. Used in fireproof materials but its dust is a health hazard.
Barite	Source of barium chemicals (as used in dyes, paints and medicines). Finely crushed barite is used as drilling mud in boreholes because of its ability to withstand pressure.
Calcite	Some used as source of lime-based chemicals or as a flux in steel making. Most calcite is extracted in the form of limestone which is used by the construction industry: e.g. cement, aggregates, etc.
Clay	Ordinary clays for bricks, tiles, etc. China clay (kaolinite) for crockery.
Fluorite	Manufacture of fluorine chemicals (e.g. hydrofluoric acid). Also as a flux in steel making.
Graphite	Soft pure carbon. Used as the 'lead' in pencils, for some electrical components and added to certain lubricants (e.g. graphite grease).
Gypsum	For plaster products. Some gypsum is also added to cement to control its rate of setting.
Halite	Common salt. For chemical processes (e.g. making caustic soda and hydrochloric acid), food industries and road 'de-icing'.
Potash	For chemical industries and manufacture of fertilisers.
Quartz	Small quantities used in electronic industries (for example, oscillators for watches) but these high-quality crystals are more likely to be 'grown' in laboratories than mined. Hardness of quartz makes it suitable for some abrasives. Quartz in the form of sand and sandstone is widely used for building purposes.
Sulphur	Can be mined in some volcanic regions or obtained by processing iron pyrite (FeS_2). Used in production of sulphuric acid, synthetic rubber, insecticides and fertilisers.
Talc	Very soft form of mica (1 on Mohs' scale). Used for talcum powder and as a filler to add bulk to some powders and pastes.
Gem stones	Precious varieties such as diamond, sapphire, ruby, opal, jade and emerald. Semi-precious types such as amethyst (quartz) and garnet.
Building stones	Wide range of uses and types, e.g. for direct building (stone blocks and slates), aggregates, road chippings, cement manufacture, etc.
Fossil fuels	Coal, oil and natural gas.

Mineral deposits

Economic minerals can be classified into two groups: **metallic** and **non-metallic**. Metallic minerals are mined for their content of metal (for example, iron from haematite). Non-metallic types are in demand by chemical companies (for example, chlorine and caustic soda are produced from halite) or as raw materials for industry (for example, gypsum to make plaster). See Table 15.1 for a full list of minerals and their uses.

Re-reading pages 19 and 20 will remind you that most of the economically important elements make up just a tiny percentage of the earth's crust: for example, copper 0.007%, uranium 0.0004%. However, these figures are only averages and such elements are not spread evenly throughout the crust. Instead, certain geological processes have concentrated them into particular places and produced deposits of minerals which are worth extracting. These concentrating processes are divided into three main types: **magmatic, hydrothermal** and **sedimentary** (see Table 15.2 on page 178). Before dealing with

each of these in more detail, we need to define some common mining terms.

- **Ore mineral**: the valuable mineral which is extracted and sold.
- **Gangue mineral**: a worthless mineral found with the ore. For example, quartz and calcite are common waste materials from many metal mines. However, what is considered to be a gangue mineral at one time may prove to be an economic ore at some later date. For example, the spoil tips of some eighteenth-century and nineteenth-century Pennine lead mines are now being worked for the barite and fluorite they contain. Today, both minerals are important in the chemical industry, but in the past they were just dumped because there was no market for them.
- **Ore body**: the total body of minerals present; this may include several ore minerals and a number of gangues.
- **Vein**: a vertical (or near vertical) body of minerals usually found infilling a fault or joint.

Magmatic deposits

These deposits are produced by economic minerals crystallising directly from a magma and becoming concentrated in a certain part of the intrusion. The best examples are associated with **gabbros**.

Gabbro magmas are relatively rich in iron and can also contain up to 200 parts per million of chromium and nickel. As cooling takes place, **magnetite, chromite** and **nickel ores** may crystallise and sink to the bottom of the intrusion forming an economic concentration of minerals.

Hydrothermal deposits

As the name suggests, these deposits are formed by hot watery fluids (hydro means water, thermal means heat). These fluids carry minerals in solution until, in suitable chemical conditions, crystallisation takes place. Before discussing the origin of hydrothermal fluids we must consider the type of deposit they produce.

(a) Normally the minerals crystallise to form veins along faults or joints; these fractures were the 'plumbing system' along which the fluids travelled.

(b) The commonest metallic minerals are sulphides of lead, zinc, copper and iron; associated with these are non-metallic minerals such as quartz, calcite and fluorite.

(c) Fig. 15.3 shows a typical specimen from a mineral vein; notice how the crystals occur in a distinctly layered pattern, this suggests minerals formed in order (one upon another) as fluids flowed along the vein over a period of time.

Hydrothermal deposits are easiest to explain where they are found close to **granite intrusions**: for example, tin and copper veins in rocks close to the Cornish batholith. A typical granite magma contains traces of a variety of economic elements such as lead, zinc, copper and barium. In addition, all magmas contain a relatively large amount of water vapour (remember that volcanic eruptions always involve the release of steam). As granites cool, neither the trace elements, nor the water, are used up by the crystallisation of the rock's normal silicate minerals (quartz, feldspar, etc.). This means that, towards the end of the cooling process, all the trace elements become concentrated in the remaining mass of water. This water moves up and out into the surrounding rocks as a mineralising hydrothermal solution.

Granite magma may not be the only source of hydrothermal solutions. In fact many veins are found well away from any known batholiths. Geologists have suggested the following theory to explain this type of deposit.

- When marine sediments are deposited they contain a large amount of salty sea water (brine).
- The brine becomes heated as the sediments are buried; temperatures of several hundred degrees centigrade may occur in the lower levels of thick sedimentary basins.
- The hot brines migrate upwards (due to pressure) and pick up traces of metallic elements from the sediments they pass through. (Brines are much more likely to react with metallic elements than pure water – think of the rate at which iron rusts in salt water compared with the rate in pure water.)
- The solutions crystallise and form mineral veins in the faults and fissures of rocks nearer to the surface.
- The upward movement of the solutions can be likened to the migration of petroleum (page 182), particularly since many veins occur in anticline structures where impermeable rocks prevented further upward movement of the fluid.
- The overall effect of this process is, over millions of years, to 'collect' traces of important elements from a vast volume of buried rock and concentrate them in veins nearer to the surface. How convenient for present-day mining companies!

One particular piece of evidence to support this theory has come from a borehole in southern California. This was drilled through a thick series of sedimentary rocks and eventually reached a point where hot, extremely salty water was flowing deep underground. This water precipitated copper onto the sides of the drill shaft.

Fig. 15.3 Layered mineralisation from a hydrothermal vein

Fig. 15.4 Abandoned underground workings on a mineral vein (note that the sloping roof of the mine tunnel is the fault plane along which hydrothermal mineralisation took place)

Sedimentary precipitated deposits

The formation of sedimentary beds of **halite, gypsum, potash** and **ironstone** depends on the right chemical conditions for their precipitation. These deposits were dealt with in the section on sedimentary rocks and you should look at pages 72 to 74 for details of their origins and uses.

Beds of **halite** within Britain's Permo–Triassic desert sediments have given rise to chemical industries in Cheshire and Teesside. The salt is not normally mined by the usual methods of underground digging but is pumped to the surface dissolved in water. Two boreholes are drilled into the halite bed; warm water is passed down one and pumped back up the other carrying dissolved salt. Vast underground caverns have been dissolved by this type of extraction and companies must be careful to avoid removing too much from one particular area. In the past, overpumping of salt beds has caused underground collapse and subsidence of the ground surface.

Britain's deposits of sedimentary ironstones such as **limonite** are generally of rather low-quality ore. Recently there has been a reduction in the amount of this material being extracted. Steelworks now tend to use better-quality imported ore and are therefore located near to ports rather than the outcrops of British ironstone.

Residual deposits

These are a result of intense **chemical weathering** (page 43) in the hot wet climates of tropical lands. As surface rocks decompose, nearly all the weathered material is removed in solution. However, hydroxides of iron and aluminium (common products from the chemical break down of silicate minerals) are very insoluble and therefore remain as a residue in the soil. Such soils are deep red in colour and known as **laterite**.

Where the original rocks contained only a very small percentage of iron, aluminium hydroxide becomes the chief **residual** (remaining) material. Such deposits are called **bauxite** and form the world's main supply of **aluminium ore**.

Placer deposits

These deposits also occur as a result of rocks breaking down at the earth's surface. In this case the action of moving water concentrates eroded fragments of economic minerals by depositing (placing) them at certain sites. A good example is the deposition of grains of **gold** in river gravels. Minerals found in placer deposits must be
- hard and resistant, to survive weathering, erosion and transportation without disintegrating in the process;
- denser than normal sedimentary material; so that they are deposited separately. For example, fragments of gold (density 19.3 g/cm³) are left by slow-moving water on a river bend while sand grains (density 2.7 g/m³) are carried further on.

Magnetite, cassiterite and **diamonds** can also be deposited by this method, and placers may develop on beaches as well as in rivers. Since they are always in loose sediment, placers are very easy to extract. In some places all you need is a shovel and a pan!

Fig. 15.5 Dredging for (placer) tin in Malaysia

Table 15.2 Formation of economic minerals

Type of process	Where formed	Ore 'body'	Main minerals	Examples
Magmatic	Mainly in large gabbro intrusions	Mineral-rich 'layer' near the base of an igneous intrusion	**Magnetite**, Fe_3O_4 **Chromite**, $FeCrO_4$ Ores of **copper** and **nickel**	Iron and chromium from mines in the Bushvelt Gabbro of South Africa
Hydrothermal	(a) Close to granite batholiths (possibly within metamorphic aureole)	Vertical or steeply dipping veins often infilling faults or joints	**Chalcopyrite**, $CuFeS_2$ **Galena**, PbS **Haematite**, Fe_2O_3 **Cassiterite**, SnO_2 **Pyrite**, FeS_2 **Quartz**, SiO_2 (gangue)	Copper and tin from mines near Cornish granite Copper* and lead* from mines near Lake District granite Copper* and lead* from mines near granites in the Southern Uplands of Scotland
	(b) Well away from granite batholiths	Vertical or steeply dipping veins often infilling faults or joints	**Galena**, PbS (**lead**) **Sphalerite**, ZnS (**zinc**) Ores of **copper** **Fluorite**, CaF_2 (**fluorine**) **Barite**, $BaSO_4$ (**barium**) **Calcite**, $CaCO_3$ (gangue) **Quartz**, SiO_2 (gangue)	Lead* from Mendip Hills Lead,* zinc,* fluorite and barite from Northern Pennines Lead,* fluorite and barite from Derbyshire Lead,* zinc* and copper from Mid Wales
Sedimentary	(a) Where a suitable sedimentary environment allowed minerals to precipitate	Bed(s) within sequence of other sediments	**Halite**, NaCl **Gypsum**, $CaSO_4.2H_2O$ **Limonite**, FeO(OH) **Siderite**, $FeCO_3$ **Potash salts**	Salt from Cheshire and Teesside (used in chemical industry) Gypsum from Nottinghamshire and Appleby (Cumbria) Ironstone* in Jurassic rocks of Northamptonshire and Yorks.
	(b) Residual deposits	Part of the surface soil	**Bauxite**	In 'soils' of Jamaica (from weathering of impure limestone)
	(c) Placer deposits	Stream banks, beaches, etc.	**Gold** **Cassiterite**	Gold in Klondike (Alaska). Tin in river muds of Malaya, etc.

* British deposit now practically worked out but worth visiting for specimens

Case study: a proposed mining development in Mid Wales

The extraction of economic minerals can have a dramatic effect on the environment. This study shows some of the problems and decisions which may face the people involved.

The hilly and rather remote area of Wales between Dolgellau and Trawsfynydd contains Lower Palaeozoic sedimentary and volcanic rocks. Within the sequence are hydrothermal deposits of copper and gold which generally take the form of small veins. Some of the gold has also become concentrated into placer deposits along rivers. During the nineteenth century many mines worked these ores but eventually they were all abandoned as uneconomic. At present only the small Clogau Gold Mine is operating and this has only been reopened quite recently (see Fig. 15.6).

During the early 1970s, interest in the region was renewed when RTZ Ltd, a multi-national mining company, resurveyed the area and discovered low-grade copper mineralisation at Capel Hermon. RTZ also considered testing for possible placer gold deposits in the Mawddach Estuary, and applied for permission to carry out both these drilling projects.

The area has superb scenery and lies within Snowdonia National Park. There was therefore some opposition to possible mining development. A public enquiry was held to consider RTZ's plans and, although permission was granted to carry out the survey programmes, RTZ decided to withdraw from both projects. Many complex questions were raised by this enquiry, and these are still relevant. The minerals are still in the ground and industry still needs copper and gold. Read through the information which follows to understand some of the important points.

Copper at Capel Hermon

The deposits here occur as a mass of very tiny 'veinlets' spread through a large volume of rock. The proportion of copper is only 0.3% which means that, on average, 1000 tonnes of rock must be extracted and processed to gain just 3 tonnes of copper metal. To make this profitable, operations need to be on a vast scale. It was suggested that RTZ would have to work the site as an **open cast** pit (at the surface) from which 350 million tonnes of rock would be mined over 15 to 20 years. This is the major difference between modern metal mining and metal mining in the past. Most of the older mines tended to work underground and extract high-quality ore from relatively small areas. Now these better-quality ores are mostly worked out. Only the lower-grade deposits are left and, to make these economic, very large-scale surface workings are needed. As a result, modern metal mining tends to have a much more damaging effect on the environment, although companies can make great efforts to restore an area once their operations are over.

Calculate the amount of copper which could have been extracted at Capel Hermon (350 million tonnes of rock to be mined, each 1000 tonnes should give 3 tonnes of copper).

How much would this copper be worth? (You can find the current price of a tonne of copper in the financial sections of some newspapers.)

What size would the opencast pit have been if 350 million tonnes of rock had been extracted? (1000 tonnes of rock has a volume of approximately 375m^3.)

Gold in the Mawddach Estuary

Many amateur prospectors have tried panning for gold in the streams around Dolgellau but as they normally keep this secret it is not really known whether anyone has ever made much money from it! RTZ's exploration plans involved the main estuary rather than the tributary streams. Had gold been discovered in commercial quantities, then workings could have been on a very large scale. In this type of operation, a barrier is sometimes built across the water, and a dredger (100 m long and 25 m high) is used to lift sediment over a period of approximately 20 years. Once any gold has been extracted, the sediment is returned to the water.

Fig. 15.6 Proposed mining developments in Mid Wales

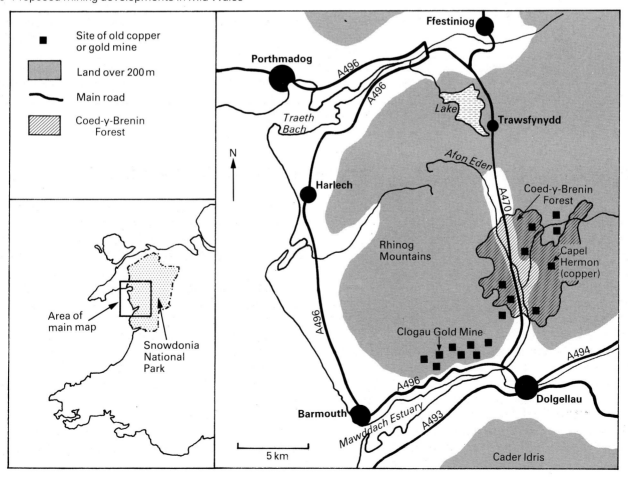

Other points to consider

- If mining at Capel Hermon and/or dredging in the Mawddach Estuary had been carried out, the developments might have involved large-scale schemes to improve power supplies and roads in the area, plus the re-routing of streams and rivers.
- It was suggested that both schemes would provide much-needed employment for this rural area.
- Many of the people who objected to the schemes stated that the beautiful environment would be ruined. However, most of these people did not live in the area but enjoyed visiting it. Should they have as much say in any decisions as the local population who actually live there?
- At present Britain imports practically all the copper and gold it needs. Would mining some of its own help the country's economy?

Having read the information try to put yourself in the position of one of the following people. Write about some of the arguments they might have put forward.

(a) A farmer whose land has belonged to the family for several hundred years but which would be needed by the mining.

(b) A conservationist who lives in an English city, but uses the area for holidays, is concerned about pollution and caring for the environment.

(c) A local person who has a low-paid job in a café during the tourist season but is unemployed in winter, has always lived in the area, and is a member of Plaid Cymru (the Welsh Nationalist Party).

(d) The local M.P. who is concerned about the national economy, local employment and keeping his/her seat at the next election.

Fig. 15.7 Area of proposed mining development in Mid Wales

Sites of old gold mines, e.g. Clogau

Coed-y-Brenin Forest

Mawddach Estuary

Energy from the rocks

Our life style depends on energy for heating, lighting, driving machinery and powering the means of transport and communications. Most of this energy comes from the earth's store of fossil fuels:

- **coal** formed from fossil vegetation;
- **oil** formed from fossil plankton;
- **natural gas** associated with coal and oil.

Coal

Before reading this section you should revise the origins and properties of different coal types (pages 71 to 72) and remind yourself how Britain's coal deposits were formed in the coastal mudflats and deltas of Upper Carboniferous times (page 162).

Most coal mines in Britain extract bituminous coal from **seams** between 1 and 2 metres thick. Thinner seams than this are generally not economic to work. Other features of a seam may also affect whether it is worth mining: for example, the depth at which it is found, its angle of dip and the number of faults. Since porous rocks such as sandstone often occur in sequences of coal-bearing rocks, many mines require expensive pumping systems to prevent flooding underground.

A typical underground mine (Fig. 15.8) has several seams, so coal is extracted at different levels, all linked by a series of shafts and tunnels. The **longwall method** of mining is used, where a power shearer and conveyor (Fig. 15.9) remove a slice from a **coal face** some 200 m long. After each complete cut along the face, the whole system of machines and hydraulic props is moved forward so that the next slice can be removed. As the props are moved, the overlying rocks press down and eventually collapse into the

Fig. 15.9 Coal cutting machinery in action (the water sprays help to reduce the amount of dust)

area where coal has been removed. Collapses of this type can cause subsidence at the ground surface with damage to buildings, roads and pipelines.

Many of Britain's older coal mines were relatively small and unmechanised collieries situated on **exposed coalfields** (places where Upper Carboniferous coal-bearing rocks outcrop at the surface). As these mines have become worked out or have met serious geological problems (faults, flooding, subsidence, etc.), there has been a trend to develop large mechanised collieries in the **concealed coalfields**. Concealed coalfields are places where younger rocks overlie (and therefore conceal) the coal-bearing beds below. Newly developed areas include Selby, Yorkshire (beneath overlying Triassic rocks) and the Vale of Belvoir, East Midlands (beneath overlying Jurassic rocks).

Fig. 15.8 Section through a coal mining area. Notice that a system of tunnels links several seams to each mine shaft. Notice too that the same seam may be worked, at different points, by several separate mines. *Refer to Fig. 14.19 on page 163. What type of sediments will be found between the coal seams?*

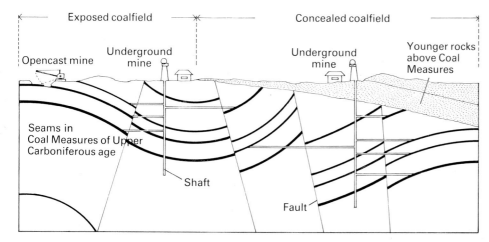

In addition to underground mining, some coal may be extracted by **open cast** methods. The overlying rock and soil (up to 80 m thick) is removed by draglines and replaced after the coal has been removed. This method is generally cheaper than underground working.

Over 60% of Britain's coal output is used in power stations to generate another form of energy: electricity.

Oil and natural gas

Sea water is rich in minute forms of life called plankton and after death these creatures tend to become incorporated in sea-floor sediments. This is particularly likely to happen in the fine muds of offshore regions where there is little oxygen and few scavengers. If conditions are favourable, these sediments become the **source rocks** from which oil and gas come. During **compaction** the remains of the plankton are slowly altered (by chemical and bacterial action) to form organic compounds called **kerogen**. If the rocks are then further buried and heated, the kerogen can break down to yield oil and gas. However, the outcome of this process can be affected in several ways.

- The kerogen-bearing sediment may not be buried deeply enough for breakdown to occur. In this situation a rock type known as **oil shale** is formed. This is mined in some parts of the world (for example, Canada). When it is crushed and heated, oil can be produced from it.
- The kerogen may be completely destroyed if it is buried deep enough for temperatures to rise above 200 °C.
- If depth and temperature are suitable for oil and gas to be formed then these will tend to **migrate** (move) away from the source rock.

Oil and gas migrate as they are carried upwards by groundwater through porous rocks such as sandstone. Movement may continue to the surface and form a natural seepage. But often the fluids fail to reach this far because they have become trapped somewhere on their upward journey. **Oil traps** can occur in a variety of geological situations (see Fig. 15.10) but the essential requirement is always an

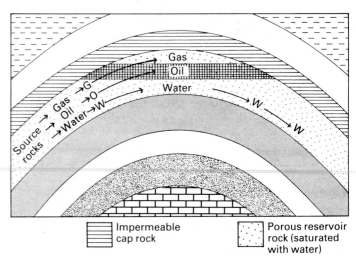

Fig. 15.11 Layered effect of gas/oil/water in trap structure

impermeable **cap rock** which prevents further upward movement. The porous rock below the cap is sometimes called the **reservoir rock**. The pores in this rock can hold water, oil or gas. Fig. 15.11 shows how a layering effect occurs within an oil trap with gas resting on oil, resting on groundwater in the rock. This is a result of the different densities of the trapped fluids (the lightest material rises to the top).

As well as coming from kerogen-bearing rocks, natural gas can also be formed during the alteration of vegetation into coal. Once again migration will carry it from its original source, but in this case there will be no oil found with it in the final trap.

Natural gas is mainly used directly as a fuel for heating. Crude oil can be separated into a range of products including road tar, grease, diesel, petrol and paraffin. It is also the raw material for plastics and artificial fibres such as nylon. Oil is therefore in great demand in industry (too precious just to burn?).

North Sea oil and gas

Large deposits of oil tend to occur where thick sequences of marine sediments have provided suitable source rocks. The best areas to survey are sedimentary basins which have never been intensely deformed by an orogeny. What is needed is enough folding to produce trap structures but not so much heat and pressure that any kerogen in the rocks was destroyed.

The North Sea Basin was thought to be a likely area because it had been subsiding since Permian times and was known to contain sedimentary rocks rich in organic material. Following the discovery of natural gas at Groningen (Netherlands) in 1959, seismic surveys were carried out across the southern North Sea to discover if any similar deposits lay offshore. These surveys revealed rock structures which could have acted as traps and the work was well rewarded when sizeable reserves of gas were proved by drilling off the East Anglian coast in the early 1960s. The

Fig. 15.10 Oil trap structures

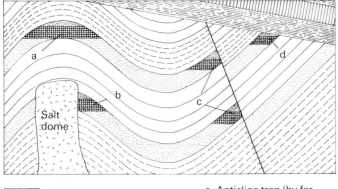

Gas	**a** Anticline trap (by far the most common: 80% of world supplies come from this type of trap)
Oil	**b** Salt dome trap
Porous reservoir rock (all other rocks are impermeable)	**c** Fault traps
	d Unconformity trap

reservoir rocks for this gas are porous sandstones of Permo–Triassic age and it is thought that Upper Carboniferous Coal Measures are the most likely source.

With proven deposits in the southern North Sea, attention then turned to the deeper waters in the north of the basin. Despite the expense and difficulty of working far offshore in stormy conditions, this northern area has provided some extremely profitable discoveries. In this region oil is the main deposit although there is also some gas. The reservoir rocks here are of Mesozoic and Tertiary age and it is believed that fine-grained sediments such as the Kimmeridge Clay (see page 167) are the most probable source.

With the successes of the North Sea in mind, other marine basins around Britain are now being surveyed for possible reserves of oil and gas. There have been discoveries in Morecambe Bay and off the coast of Cork in Southern Ireland. Other areas look promising.

Fig. 15.13 North Sea oil platform

Other geological power sources

Geologists are involved in the discovery and mining of **uranium** ore for nuclear power stations, and are also searching for (safe?) underground storage sites for radioactive waste.

Water heated by volcanic action provides steam for **geo-thermal** power stations in countries such as Italy and New Zealand.

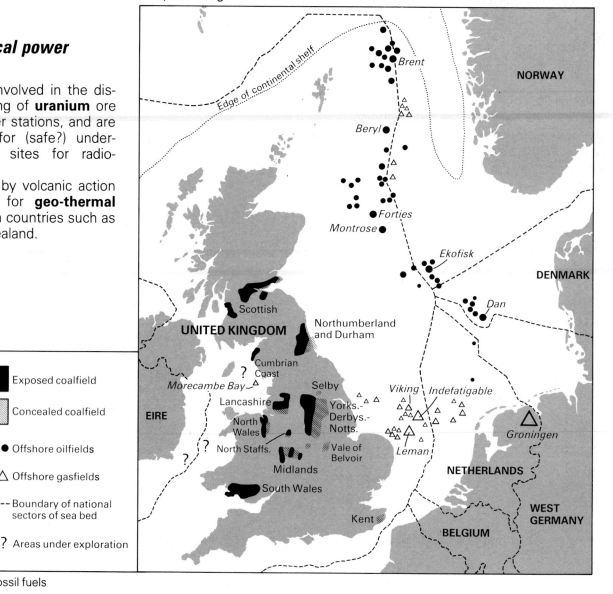

Fig. 15.12 Britain's fossil fuels

183

Geology and the construction industry

Building stone

In the days before cheap mass-produced bricks were available, many buildings were made of solid stone. Since transport was expensive, each area used the most suitable local rocks. In general the harder types of sedimentary rock were preferred because bedding planes and joints made quarrying and cutting easier. Examples of local building stones are Carboniferous limestone in the Pennines, Carboniferous sandstone (such as Millstone Grit) in the older industrial towns of Lancashire and Yorkshire, Jurassic limestone in the Cotswolds, and Triassic sandstone in Cheshire.

Local rock was also used in stone walls built to enclose farm land in the late eighteenth and early nineteenth centuries. You can often tell where geological boundaries occur simply by noticing where the type of rock changes in these walls.

Best-quality stone is carefully cut and/or polished for use in important buildings such as town halls, banks and monuments. *How many rocks can you identify in the centre of your town? Look out in particular for marbles and igneous rocks. Many of these may be imported (for example, Italian marble), since only a few British quarries now produce polished stone. Don't forget to look upwards; your survey would not be complete without a note on the types of roofing slate.*

Aggregates, chippings and sand

These three products are the most widely quarried materials in Britain with hundreds of millions of tonnes being extracted every year.

Aggregates are stone fragments. They are used either directly for infilling on construction sites (for example, foundations of roads and houses) or are mixed with cement to make concrete. Chippings is the common name for the small angular pieces of rock that are rolled into road tar to improve tyre grip.

Most of the harder rock types can be used as aggregates or chippings; they are simply crushed and sorted into different sizes at the quarry. To avoid high transport costs aggregates must be locally supplied and each area tends to use what is available: for example, granite in south west England, limestone in the Pennines, and dolerite (from the Whin Sill) in north east England. Most aggregate quarries use road transport so that their products can be delivered directly to construction sites. Since the quarries tend to be in rural areas, large numbers of heavy lorries are often forced to travel on narrow roads and through small villages to reach them. Quarry noise and dust from the blasting and crushing can also cause annoyance. But these disadvantages must be balanced against the employment that quarries provide in an area. There have been particular objections to the amount of quarrying which takes place in Britain's National Parks, but to some extent this cannot be avoided because the most suitable hard rocks tend to occur in the rural uplands of the country.

Fig. 15.14

Some parts of lowland Britain have 'drift' deposits of **gravel** laid down by rivers or glacial meltwater. These are easily dug out and, when washed to remove clay particles, make excellent aggregates. Similar surface deposits of loose **sand** are also excavated where available. Some low-lying pits are water filled and worked by dredging. These can sometimes be landscaped and used for water sports when production ends: for example, Thorpe Water Park at Chertsey, London.

Although beaches seem obvious places for the collection of sand and gravel, the salt content weakens any concrete made with such material.

Glass sands

Very pure quartz sand is 'melted' to make glass. For example, Shirdley Hill Sand (a Quaternary age deposit of wind-blown sand) supplies the many glassworks at St Helens, Lancashire.

Cement

This vital building material is made from **limestone** plus **shale** or **clay**. The rocks are crushed, mixed together and burned in a kiln at over 1000 °C. Chemical reactions take place and when the substance has cooled it is ground into a fine grey powder.

A typical cement works location is shown in Fig. 15.15; note the importance of good transport links. Similar sites exist elsewhere on Carboniferous limestone, Jurassic limestone and Cretaceous chalk (provided shale or clay is also available nearby).

Fig. 15.15 Site of cement works, Castleton (Derbyshire)

Clay and china clay

As Fig. 7.11 (page 69) illustrated, the main economic use of clay is for making **bricks**. For this purpose it is important that the clay does not contain too much sand or the bricks will fail to harden properly when fired in a kiln. The colour and texture of British bricks varies according to the type of local clay used but the chief area of production is between Bedford and Peterborough, where the Jurassic **Oxford Clay** occurs. In addition to bricks, clay is also used in the manufacture of tiles, drainage pipes, etc.

Only the best pure white **china clay** is suitable for making crockery and porcelain. Such clays are made of the mineral **kaolinite** and are mined, for example, from altered granite in the outer zones of the Cornish batholith. Kaolinite formed during the late stages of the granite's cooling when hot watery fluids passed through the rock and caused chemical alteration of feldspar. The reactions for this alteration are almost identical to those which occur when feldspars are broken down to clay minerals during chemical weathering (see page 43). Alteration of feldspars has left parts of the Cornish granite so weak that it can be broken up in the china clay pits by jets of water from hoses. The water washes away the fragments of rock into settling tanks and here the kaolinite can be separated because it remains in suspension while the coarser pieces of waste quartz and mica sink to the bottom.

Plaster

This material is produced by heating and evaporating the water from crushed gypsum. In Britain deposits of Permian age are mined for this purpose, for example near Appleby in Cumbria.

Fig. 15.16 Geological building material

Water Supplies

Underground

Beds of rock which provide a supply of underground water are called **aquifers**. As explained on page 50, such rocks must be either **porous** (for example, sandstones) or **permeable** (for example, limestones).

In general, sandstones make the best aquifers, especially if they are poorly cemented and/or contain only a small amount of clay matrix. (Without a cement or matrix there are plenty of pore spaces between the grains for water to collect in or pass through). In Britain, **wells** have been drilled into a variety of sandstones of different ages. An interesting example is at Burton-on-Trent where the town's many breweries originally depended on the 'taste and chemistry' of well waters from Triassic sandstones.

Water in an aquifer fills it to a level called the **water table**. As Fig. 15.17 shows, water tables vary according to the amount of rainfall. They can also be affected by the rate of pumping from wells.

Some aquifers hold water under pressure so it rises from wells without the aid of pumps; the **artesian** basin formed by the Chalk syncline beneath London is an example (see Fig. 15.18).

Throughout the country, local supplies are also obtained from **springs**; some is even bottled and sold as spring water.

Surface reservoirs

Impermeable bed rocks are obviously needed for a reservoir site. The only possible exception to this is where a thick covering of clay drift provides a seal between the water and a porous or permeable bed beneath. Geologists must also make certain that the underlying rock structures are suitable for the dam's foundations and the weight of water to be supported. The building of one particular reservoir in India caused a minor earthquake as rocks adjusted to the pressure.

Fig. 15.17 Wells and water table. *What happens to supplies from well B during a drought? How could the well be modified to overcome this problem (there are two possible ways)?*

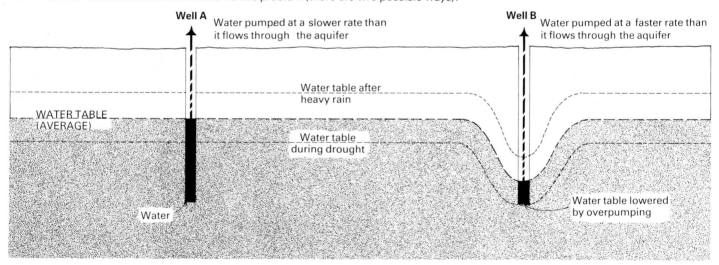

Fig. 15.18 Artesian basin of the Chalk syncline beneath London

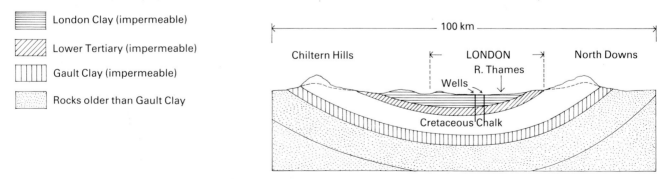

--------- Water table in Chalk. This used to be above the surface level of the London wells so that water rose from them without pumping. However, to increase the supply, pumps were installed and as a result the water table has now been lowered.

The future of mineral resources

Land will grow crops year after year, water supplies are replaced by rainfall but, when a deposit of rock, mineral or fossil fuel is used up, all that is left is a hole in the ground. In other words, geological deposits are **finite** and **exhaustible** (finite means there is a fixed limit to their amount; exhaustible means that it is possible to use them up completely).

The demand for geological materials and fuels has increased dramatically during this century. This demand is likely to rise even faster as developing countries try to improve their living standards and set up their own manufacturing industries. Supplying future demands depends on the ability of geologists to keep finding more and more deposits to replace those that have been already used. Obviously such discoveries cannot go on for ever but it is nevertheless impossible to predict when (and even if) particular materials will eventually run out. Provided geologists keep developing more sophisticated exploration techniques, there is always a chance of finding deposits that previous (less accurate) surveys have missed. Of course, if a deposit needs really advanced techniques to find it, it is also likely to be more difficult and expensive to extract (for example, it could be buried more deeply or contain a lower percentage of the required material than the more obvious deposits found by older methods). In some ways geological exploration is rather like searching in a haystack: once you have found and used all the easy things only the 'needles' remain. Since these 'needles' are your only resource you have no alternative but to improve your search methods until you find them!

Improved survey methods may also allow previously unreachable areas to be studied. This is particularly true in the oil industry which has developed the technology to find and extract offshore deposits. The rocks of continental shelves may well contain reserves of other minerals but developing methods of underwater mining will be a massive challenge. For the very distant future we might even consider getting some of our materials from other planets; then as now, it would be a matter of whether the cost of obtaining them was less than the price they could be sold for.

In addition to continually searching out new deposits, two other important steps can be taken to ensure that our resources last as long as possible. Firstly, alternative materials need to be developed which can take the place of those in short supply (for example, there is an increasing use of plastic water pipes instead of copper). Secondly, far more material should be re-cycled rather than dumped (there is a need here to develop mechanised methods of breaking down complex machines into their component materials). Unfortunately it is not possible to re-cycle fossil fuels: once coal, oil and gas are burnt they have gone for ever. The only way to ensure that we are not one day left without power is by developing alternative supplies which can never run out: for example, using nuclear, solar, wind, tide or wave power.

Questions

For questions 1–5 write the letters (A, B, C, ...) of the correct answers in your notebook.

1 River gravels containing grains of gold are an example of
   A residual deposits
   B evaporite deposits
   C placer deposits
   D magmatic deposits
   E hydrothermal deposits

2 Which **one** of the following ores provides a metal used to coat the inside of food cans?
   A azurite   B cassiterite   C galena
   D graphite   E pitchblende

3 Which **two** of the following mineral ores are most likely to be found associated with large intrusions of gabbro?
   A chromite   B limonite   C sphalerite
   D malachite   E magnetite

4 The map shows the stream pattern of an area where a geochemical survey has been carried out in search of galena. The dots represent places where the metal content of the water (dissolved from galena) has been analysed and measured in parts per billion.
   At which of the sites (A to E) would you expect the galena deposit to be?

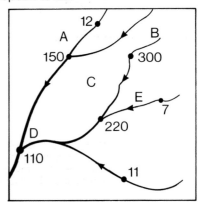

187

5 Which **one** of the following minerals is used in the manufacture of plaster?

   A halite   *B* gypsum   *C* amphibole
   D fluorite   *E* olivine

6 Study this geological cross section through a coal mining region. Points A to E are possible sites for coal mines.

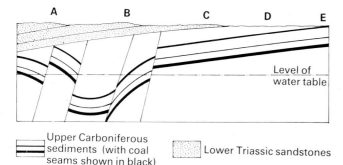

   Upper Carboniferous sediments (with coal seams shown in black)   Lower Triassic sandstones

(a) Which mine sites are on an exposed coalfield and which are on a concealed coalfield? (Explain your answer.)

(b) Carefully explain why mining at sites A and B would be more difficult and expensive than at any of the other sites.

(c) Which site would be most suitable for an open-cast coal mine?

(d) During which period of time are the faults most likely to have developed?

(e) The coal seams are found within repeated sequences of deltaic sediments (cyclothems); draw a labelled diagram to illustrate a typical cyclothem and explain what type of conditions could have produced it.

7 The map below shows the location of a mineral vein and the table lists the properties of the three main minerals found in it.

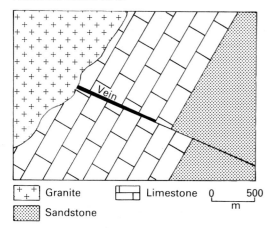

   Granite   Limestone   0   500
   Sandstone   m

Mineral	Hardness	S. G.	Colour
A	7	2.7	glassy white
B	4	3.1	purple
C	2.5	7.5	silvery grey

(a) Name minerals A, B and C and state their chemical composition.

(b) What is the meaning of the term gangue mineral; which of the three minerals (A, B or C) would you describe as such?

(c) Using information from the map carefully explain how this vein may have formed.

(d) Assuming the vein to be 2 m wide and 125 m deep calculate the total volume of this ore body.

(e) If each cubic metre of this ore body yields, on average, 0.1 tonnes of mineral C, what weight of this mineral could be obtained if the entire vein was mined?

8 The diagram below shows a cross-section through an area where it is proposed to build a new reservoir. What do you think would happen if the reservoir was built and filled? Give reasons for your answer.

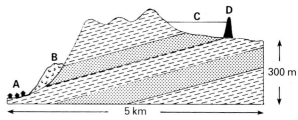

   A Large housing estate
   B Spoil heap from old mine workings
   C Proposed reservoir water level
   D Proposed site of dam

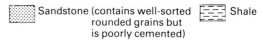

   Sandstone (contains well-sorted rounded grains but is poorly cemented)   Shale

9 Draw a diagram to show the movement of water in an artesian basin. Name an artesian basin which is used for water supplies in Britain.

10 Name one important mineral deposit in the British Isles. State the properties of the mineral concerned and explain how this particular deposit could have formed. What industries rely on it?

11 Write an illustrated essay on **one** of the following topics. In each case a list of terms is provided to help you.

(a) The origin and migration of oil (cap rock, source rock, kerogen, anticline, porous, oil trap, plankton, natural gas, offshore marine sediments, burial, reservoir rock, groundwater, oil shale)

(b) The origin and migration of hydrothermal fluids (vein, magma, faults, ore minerals, solution, joints, brine, sulphides, gangue minerals, crystallisation, batholith)

12 A company has applied for permission to open a large limestone quarry and build a cement works in part of the Yorkshire Dales National Park. Explain the effects this could have on the environment.

Appendix 1: Fieldwork

Fieldwork is a vital part of learning about geology. It provides your only chance to see, examine and interpret *real* minerals, rocks, fossils and structures exactly as they are found in *real* geological situations.

Of course, you can only learn the techniques of fieldwork by going on fieldtrips, but here are a few points to consider before you start out.

Don't be a hammer-hooligan!

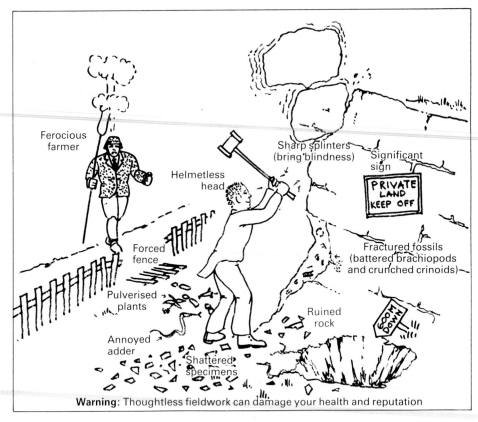

Ferocious farmer

Helmetless head

Sharp splinters (bring blindness)

Significant sign

PRIVATE LAND KEEP OFF

Forced fence

Fractured fossils (battered brachiopods and crunched crinoids)

Pulverised plants

Ruined rock

Annoyed adder

Shattered specimens

Warning: Thoughtless fieldwork can damage your health and reputation

What you need (in order of importance)

- The right attitude: go with the intention of looking, thinking and learning.
- Suitable clothing: you won't learn much if you are cold, wet or injured. Remember also to wear a safety helmet if there is any possible danger of rocks falling from above.
- A notebook, pens and pencils: you are bound to forget names, features and facts if you don't note them down *at the time.*
- A map: you should also have the ability to 'read' the map because it does help to know exactly where you are!
- A hand lens: vital for a close look (wear it on a cord round your neck).
- Measuring equipment: small tape measure, clinometer and magnetic compass.
- A supply of plastic bags and labels: there is no point in collecting specimens if they then get mixed up and you don't know where each is from.
- A geological hammer: this may not be needed at all, since at most localities there is plenty of loose material already lying around. If you do need to hammer off a fresh specimen, protect your eyes with safety goggles, warn others to stand clear and only strike the rock at what appears to be a weak point.

Before setting out
- Why am I going? What do I hope to learn?
- Have I the correct equipment?
- Have I permission to go where I want?
- Is the weather suitable? (Check forecast.)
- Have I told someone where I am going?

At each locality
- Exactly where am I? Can I give a grid reference for this spot?
- Is it safe to be here?
- Have I permission to be here?
- What am I interested in here? What can I see? What can I measure, study and note? What can I LEARN here?
- Will any of my actions damage this site?

The thinking helmet's fieldwork checklist

Field notebooks

There is little point in making rough notes in the field and writing them up later. If notes are rough, they are likely to be inaccurate, and it is difficult to correct them afterwards when you have left the site. Note everything down while you can actually see it!

During the course of a trip, you should give the name and grid reference of each locality visited. Many localities are best described by means of a simple but well-labelled sketch. Such sketches do not have to be works of art but do have to show the relevant geological features.

Look at the two example pages from a field notebook shown below. The first set of field notes shows how a detailed sketch can be used for a small locality with specific features. Note the use of sketches within sketches to highlight the really important aspects. The second shows how you can finish off a piece of fieldwork with a sketch map of the area and a list of all the localities that you visited.

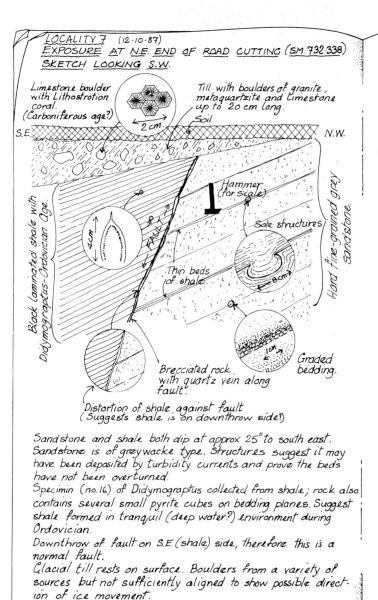

LOCALITY 7 (12·10·87)
EXPOSURE AT N.E. END OF ROAD CUTTING (SM. 732·338)
SKETCH LOOKING S.W.

Limestone boulder with Lithostrotion coral. (Carboniferous age?)
2 cm.
Till with boulders of granite, metaquartzite and limestone up to 20 cm long.
Soil
S.E. N.W.
Black laminated shale with Didymograptus-Ordovician Age.
4 cm
FAULT
Hammer (for scale).
Sole structures
Thin beds of shale
8 cm
Hard fine-grained grey sandstone.
Brecciated rock with quartz vein along fault.
10 cm
Graded bedding.
Distortion of shale against fault (Suggests shale is on downthrow side?)

Sandstone and shale both dip at approx 25° to south east. Sandstone is of greywacke type. Structures suggest it may have been deposited by turbidity currents and prove the beds have not been overturned.
Specimin (no.16) of Didymograptus collected from shale; rock also contains several small pyrite cubes on bedding planes. Suggest shale formed in tranquil (deep water?) environment during Ordovician.
Downthrow of fault on S.E. (shale) side, therefore this is a normal fault.
Glacial till rests on surface. Boulders from a variety of sources but not sufficiently aligned to show possible direction of ice movement.

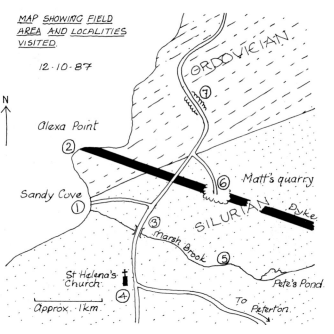

MAP SHOWING FIELD AREA AND LOCALITIES VISITED.

12·10·87

N

ORDOVICIAN

Alexa Point
②
⑦
Sandy Cove
①
⑥ Matt's quarry
SILURIAN Dyke.
③
Marsh Brook
⑤
St Helena's Church
④
Pete's Pond.
Approx. 1km
To Peterton.

① Sandy Cove (722·328) Ordovician-Silurian boundary.
② Alexa Point (720·333) Dolerite dyke in Ordovician seds.
③ Stream bed beneath bridge (726·326) Folded Silurian sediments
④ St Helena's Church (725·320) Polished marble & granite gravestones.
⑤ Stream bank (736·322) Glacial till.
⑥ Matts quarry (735·329) Dolerite dyke in Silurian seds.
⑦ Road cutting (733·338) Ordovician sediments and fault.
   (N.B. This is the locality described on the opposite page.)

Appendix 2: Ideas for practical work

As explained at the start of this book, geological study cannot be completed by just reading about the subject. This section therefore provides ideas for learning in more practical ways. There are fieldwork tasks, laboratory experiments and demonstrations relevant to each chapter of the book, plus some suggestions for projects and discussions. Have fun with these — you will improve your understanding of geology!

Chapter 2   Planet Earth

(a) Compare spacecraft photographs of the earth with those of other planets. Look for evidence of atmospheres, similar or different geological processes. Try to work out the scale of features. (Moon-earth comparisons are particularly useful.)

(b) Investigate how materials separate and form layers according to their density: for example, water, oil, air in a jar. (This can be likened to the zonation of the earth.)

(c) Investigate the effects of stress by bending, deforming and breaking a variety of materials (for example, paper, Plasticine, wooden rulers, etc.) Notice how some materials return to their original shape when the stress is removed (elastic deformation), others remain in their deformed shape and state (plastic deformation) and some break suddenly if too much stress is applied (brittle fracture). Notice too that energy (shock waves) is released during brittle fracture. What does this tell you about the type of rocks and the conditions of stress needed to produce earthquakes at various depths within the earth?

(d) Investigate how the velocity of sound waves varies in air, water and solids. This can be likened to the variation in the velocity of shock waves as they travel through different zones of earth.

(e) Make a simple seismograph (see Fig. 2.10 but use a pen instead of a beam of light.) Gently shake the table beneath your machine; is it sensitive enough to record the pattern of vibration?

(f) Calculate the density of specimens of granite and basalt (use the method given for mineral density on page 23). What does this prove about the relative densities of continental and oceanic crust?

(g) Study maps showing the topography of the sea floor. Find the plate boundaries. Describe and measure the surface features associated with them.

(h) Keep a wall map of the world and plot earthquakes and volcanic eruptions as they are reported in the news (especially useful if added to over several years).

(i) Investigate the mild radioactivity of some rocks with a geiger counter (for example, granite). Could this be the source of the earth's internal heat?

(j) Make the apparatus shown in Fig. 2.22 to demonstrate isostasy. To prevent the blocks turning over, small weights can be fixed to their bases or holes drilled so that they can slide up and down vertical wires.

Experimental and practical studies, whether in the laboratory or field, may involve potential dangers. The attention of teachers is drawn to the advice and instructions given in such documents as:

- *Safety in Outdoor Pursuits* D.E.S. Safety Series No. 2 (H.M.S.O.);
- *Safety in Science Laboratories* D.E.S. Safety Series No. 2 (H.M.S.O.);
- *Safeguards in the School Laboratory* Association for Science Education;
- *A Code for Geological Fieldwork* Geologists' Association.

Chapter 3   Minerals

(a) Investigate how elements may chemically combine to form compounds. For example, iron (Fe) + sulphur (S) → FeS.

(b) Investigate crystal growth using the method given on page 21.

(c) Make a simple goniometer (from a protractor) and measure the angles between faces of mineral crystals.

(d) Get practical experience of conducting all the mineral tests described on pages 22 to 24. Make sure your results are accurately recorded.

(e) Test, describe and identify specimens of all the minerals listed in the table on page 25.

(f) Get practical experience of identifying minerals in the field.

(g) Draw simple flow charts to show how a limited series of key tests may be used to distinguish a set of named minerals one from another. If you know how, turn your flow chart into a computer program.

Chapter 4   An introduction to rocks

Specific ideas are given (below) for each of the three main rock groups. Whenever you are studying rock specimens you must use a logical step-by-step method of observing and testing their properties. Consider your work as a 3 × 3 approach. By studying **mineralogy, texture** and **structure** you will get the information needed to **describe, identify** and **interpret**.

Chapter 5   Igneous rocks and volcanoes

(a) Investigate the relationship between lava viscosity and temperature by using other viscous substances such as treacle or tar. Add small elongated objects (for example, panel pins) to show flow banding and introduce bubbles (via straw) to produce vesicles.

(b) Experiment with 'pop bottle eruptions' to prove that gas may be contained in a liquid under pressure, and see what happens when that pressure is suddenly released! Add a few solids to the pop (peas, apple pips etc.) to show how the distance pyroclastic rock fragments can travel depends on their size. By using a corked bottle you can even have a 'lava spine' rising from the 'vent' before the inevitable blast of liquid follows – but be careful!

(c) Fill some balloons with water and stack them one on top of the other to investigate the type of structures formed when pillow lavas pile up.

(d) Look at sills, dykes, lava flows and batholiths in the field. Note their size and their relationship to surrounding rocks. Study examples of baked and chilled margins.

(e) Describe and identify hand specimens of igneous rocks (in the laboratory and field) using the techniques explained on pages 39 and 40. The use of a hand lens is essential.

(f) Study and describe thin sections of igneous rocks (use photographs or scaled drawings if a microscope is not available: see Fig. 5.17, for example).

(g) Make a large-scale version of Fig. 5.19 with real specimens placed in appropriate positions.

(h) Classify a set of igneous specimens according to their colour and then measure the density of each. Plot your results on a scatter graph like this.

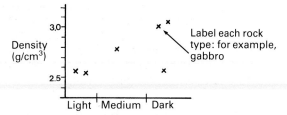

(d) Compare the colour, appearance and 'strength' of weathered and non-weathered specimens of the same rock type. This work is best carried out in the field.

(e) Demonstrate the action of abrasion by rubbing grinding paste on a glass surface.

(f) Use different mixes of sediment and stir them into beakers of water. For each mix, measure the time it takes for all sediment to settle and note the final arrangement of the grains on the bottom.

(g) Investigate the movement of sediment by water in both the field and laboratory. Fieldwork can involve measuring the speed of flow and size of deposited sediment at a number of sites along a river. Laboratory work could involve setting up your own water channel shapes (with plastic guttering?) and varying the sediment load and speed of water flow. You could try to reproduce the conditions which cause deposition in alluvial fans, deltas, flood plains etc. Make sure your studies include the movement of water around a bend.

(h) Visit field localities to study the erosional and depositional features (and sediments) produced by the actions of wind, water and ice.

(i) Look at photographs of various landscapes (for example, on postcards). Try to work out what processes of erosion and deposition have produced the scenery of that area. Perhaps you can also work out what type of bedrock is present?

(j) Conduct 'sandcastle' experiments to test the cohesive properties and angle of rest of various sediments. Use this method.

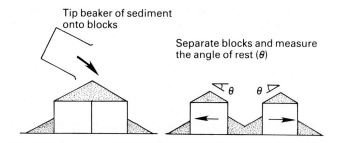

Repeat the experiment with different sediments (for example, small pebbles, sands, clays and mixtures of these materials). In each case measure the angle of rest for dry sediment and the angle for wet sediment.

(k) Make a survey of a deposit of beach pebbles. See if there are any relationships between the size and roundness of pebbles and their position on the beach.

(l) Investigate the porosity of sands by the methods given in part (m) of the study ideas listed for Chapter 15 Economic geology.

(m) Investigate transportation and sorting by wind action. See page 62 and Fig. 6.42 for details.

Chapter 6   Sediments and surface processes

(a) Investigate freeze – thaw action by the methods given on page 43. You can also measure the expansion of ice by placing a measured amount of water in a plastic syringe, sealing the 'needle' end and seeing how far the plunger is pushed back when the water is frozen.

(b) Investigate the effect of repeated heating and cooling of rocks (exfoliation) by the methods given on page 43.

(c) Investigate the acidity of water by testing samples of distilled, tap and rain water with litmus paper (this paper turns red in acid). Try to make weak carbonic acid by the following method.

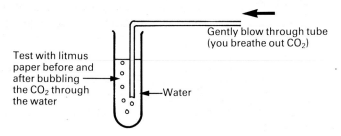

When you have made some carbonic acid, add a few limestone chippings to the tube and place a stopper in the top. Make up another stoppered tube with limestone chippings and distilled water. Compare the contents of both tubes after several weeks.

Chapter 7   Sedimentary rocks

(a) Describe and identify hand specimens of sedimentary rocks (in the laboratory and field) using the information on pages 64 to 78. Use of a hand lens is essential and descriptions should always include comments about the type of environment in which that sediment may have been deposited.

(b) Study and describe thin sections of sedimentary rocks (use photographs or scaled drawings if necessary). This work is especially useful when studying the textures of sandstone types (see Fig. 7.8) and limestones (Fig. 7.15).

(c) Investigate grain size and sorting of sands with sets of sieves. Page 67 gives details.

(d) Investigate the mineralogy of sands as explained on page 68.

(e) Try to completely dissolve samples of (i) crushed chalk and (ii) crushed muddy limestone in dilute hydrochloric acid. What does the result of this test tell you about the proportion of calcite in each of these limestone types?

(f) Investigate the burning properties (for example, flame type, heat output, ash content) of equal weights of peat, lignite, bituminous coal and anthracite.

(g) Evaporate a sample of sea water and investigate the precipitate formed.

(h) Produce 'beds' of sediment by adding pulses of material into a beaker of water; that is, tip in one lot of sediment and allow to settle before tipping in another lot (repeat as many times as you like). Each pulse should settle out to produce an individual bed but see what happens if you (i) use exactly the same type of sediment for each pulse and (ii) vary the type and grain size of each pulse.

(i) Try to produce ripples by blowing a current of air across a tray of sand (see page 75). How is the orientation and size of the ripple related to (i) the direction of the current, (ii) the velocity of the current (iii) the type of sand and (iv) the moisture content of the sand? You should also study and measure (amplitude and wave-length) examples of present-day beach or river bed ripples.

(j) Produce graded bedding and turbidity currents by the methods explained on page 76.

(k) Investigate shrinkage and the formation of dessication cracks by the methods given on page 77.

(l) Test the 'strength' of the cement in various sandstones by trying to crush them into their component grains. The presence of calcite cements can also be tested with acid.

(m) Make your own 'sedimentary rocks' by cementing sands and gravels with a slurry made from Polyfilla and water.

(n) Study a sequence of sedimentary rocks in the field and try to work out how the environment of deposition was changing as time passed.

Chapter 8   Metamorphic rocks

(a) Describe and identify hand specimens of metamorphic rocks (in the laboratory and field) using the techniques given on pages 82 to 85. Work should include (i) comparing specimens with their non-metamorphic equivalents (see Fig. 8.5) and (ii) comparing sequences of specimens which show progressive grades of metamorphism. Use of a hand lens is essential.

(b) Study and describe thin sections of metamorphic rocks (use photographs or scaled drawings if necessary: see Fig. 8.10 for example).

(c) Study a metamorphic aureole in the field and note how the intensity of metamorphism varies with distance from the heat source.

(d) Study a field area of regionally metamorphosed rocks to discover the relationships between cleavage, folding, foliation and the direction of applied stress.

(e) Study a field example of a fault breccia.

(f) Compare the formulae of minerals in metamorphic rocks with those of minerals in their non-metamorphic equivalents. Try to demonstrate that the same elements are involved and nothing is added or taken away during the process of metamorphism.

Chapter 9   Geological time

(a) Imitate the decay of radioactive elements by tossing a large number of coins. Remove any 'heads' and count how many 'tails' are left. Repeat the process until only a few coins are left. The discarded 'head' coins represent atoms that have decayed. Plot your results on a graph using the vertical axis to show how many 'tails' are left after each throw and the horizontal axis to record the number of throws. Compare the graph with the table in Fig. 9.6.

(b) The techniques of relative dating are best demonstrated in the field but map and section work (see Chapter 11) will also provide practical experience.

(c) Redraw Fig. 9.1 to compare geological time with a 24-hour period instead of a year. This means each minute would be equivalent to 3.2 million years.

Chapter 10   Earth force: deformation of rocks

(a) Investigate the effects of stress on a range of solid materials to produce elastic deformation (return to original shape), plastic deformation (folding) and brittle fracture (faulting). Vary the amount of stress, the direction of the stress (compression or tension) and the length of time that it is applied for. See also part (c) of the study ideas for Chapter 2.

(b) Use paper, Plasticine, wood blocks, etc. to model fold and fault types.

(c) Construct a simple clinometer (see page 94).

(d) Measure dip and strike in the field using a clinometer and magnetic compass. Record your results accurately (including a grid reference of the locality) and try plotting the measurements on a base map. Note that practice dip and strike measurements can be made on plane surfaces set up in the laboratory at different attitudes and orientations.

(e) Accurately measure a fold structure in the field and then make a scaled drawing of it on squared paper.

(f) Use a magnetic compass to measure a set of joint directions. This is best done on the crest of an exposed anticline.

(g) Investigate a field example of an unconformity. Particularly look for evidence of an original erosion surface and its subsequent burial by renewed sedimentation (for example, basal conglomerate). Measure the dip and strike of rocks above and below the unconformity.

(h) Calculate the true, apparent and vertical thickness of beds seen at field localities or in maps and sections.

Chapter 11 Geological mapwork

(a) Interpret simple geological maps by drawing sections and working out their geological histories. See pages 104 to 112 for details.

(b) Make 'box' models (basic outline shown below) to help your interpretation of three-dimensional structures. These models will be similar to the one shown in Fig. 11.1 but, of course, the top (map) surface will have to be absolutely flat. Either invent and draw a simple geological map on the top surface of the block or begin by drawing a section view on one of the sides (S). Then try to continue all the geological boundaries so they make a logical three-dimensional pattern of geology right round the box. When you first try this it is easier to assemble the box before adding any rock boundaries but, as you become more skilled, see if you can complete the 'geology' before the box is even cut from the paper. Do all the boundaries match up properly?

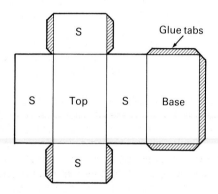

(c) Collect field evidence and try to draw your own geological map of a small area.

(d) Study British Geological Survey maps at various scales. You should look particularly at those of field areas which you have visited or which show the region near your school.

Chapter 12 Fossils

(a) Compare fossils with non-fossilised material: for example, a gastropod fossil with an empty 'modern' snail shell and a living snail. What is preserved, lost or altered by the processes of fossilisation?

(b) Draw a simple food chain from knowledge of your local environment. Discuss and estimate the chances of preservation of each member. Could any members of the chain possibly leave trace fossils?

(c) Draw, label, describe and identify representative specimens of the fossil groups dealt with on pages 119 to 136. The technique explained on page 137 will help you to draw specimens. Each description should include details of age and probable mode of life.

(d) Make moulds (latex) and casts (plaster of Paris) of fossil specimens.

(e) Study the school collection to see if particular fossil types are associated with certain rocks. Plot results in this way:

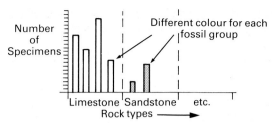

How might the graphs differ if collections from all over the country were analysed?

(f) Study some biology books and make a list of present-day animals and plants which only live in specific environments (for example, polar bears only live in the Arctic). Think about each example. Is it likely that this type of organism will ever become fossilised? What type of sediment could preserve it? If its fossil remains were ever found by geologists of the future, do you think that they would be able to work out what type of environment the original organism lived in?

(g) Study museum fossil collections (especially exhibitions of vertebrates).

(h) Study a collection of fossil specimens and decide in each case whether the organism is likely to have been preserved in its original environment or whether it was transported and redeposited after death.

(i) Make a statistical analysis of the fossil types found (not removed) at a field locality or area. Plot a histogram to show the numbers of each type present. The analysis could also involve measuring each specimen to discover if all members of the same group are of a similar size or if there are wide variations. What can be concluded from the results?

(j) Produce scaled photographs of fossils seen on field visits rather than collecting actual specimens. Add suitable captions to each: for example, grid reference of locality, age and type of rock, etc.

Chapter 13 Plate tectonics

(a) Make a model to illustrate sea-floor spreading and magnetic reversal patterns (like those shown in Fig. 13.5).

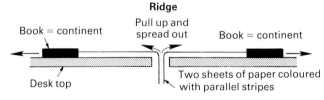

The model can be adjusted to show movement on transform faults (use two sets of paper spreading at different rates from the same 'ridge') or operated to involve subduction.

(b) Cut out continental shapes from an old atlas and try to reproduce Fig. 13.15 (Gondwanaland). Also try with continental shapes modelled in Plasticine and moved around the surface of a globe.

(c) Investigate the movement of convection currents as explained on page 148 (Fig. 13.17).

(d) Investigate why sea level may change by an experiment with water in a plastic bucket. The level can be altered by (i) temporarily removing water (ice age), (ii) putting the water back (interglacial), (iii) adding a brick to the bottom of the bucket (displacement by a spreading ridge) and (iv) distorting the bucket (plate movement?).

(e) Check through all your fieldwork notes. What evidence have you personally seen which suggests Britain has (i) been situated at different latitudes during different periods of time (proof of continental movement) and (ii) been subjected to periods of mountain building, regional metamorphism and igneous activity (proof of contact with ancient plate boundaries)?

(f) See (h) and (i) of the study ideas listed for Chapter 2 Planet Earth.

Chapter 14   The geological history of the British Isles

Field localities are the best aids to understanding aspects of Britain's geological history but laboratory demonstrations (map and specimens) are the only feasible method for more distant regions.

Chapter 15   Economic geology

(a) Identify and describe specimens of the economic ore minerals and rocks listed in Table 15.1 a and b (pages 174–5).

(b) Visit working quarries and mines. You could set up an on-going survey of one particular site which includes (for example) yearly photographs taken from the same fixed points, yearly comments from local residents, yearly traffic surveys of vehicles leaving the works, etc.

(c) Investigate how metals can be smelted from ores by using charcoal block techniques.

(d) Investigate how minerals can be separated from each other according to their densities. Experiment by 'panning' a mixture of sand and lead shot.

(e) Obtain and plot figures for annual energy consumption in Britain. If possible these should be subdivided to show relative amounts produced by different sources.

(f) Investigate the burning properties (for example, flame type, heat output, ash content) of equal weights of peat, lignite, bituminous coal and anthracite.

(g) Investigate the effect of cap and reservoir rocks with an apparatus like the one shown below. The slab of clay acts as the cap rock, and the sand acts as the porous reservoir rock. Make sure the edges (E) of the clay fit tightly against the container. In pressing the clay down it may be possible to form a slight dome at the surface: so much the better. Open tap A and allow oil to saturate about half the sand before opening tap B. The well should expel first air (gas) then oil then water.

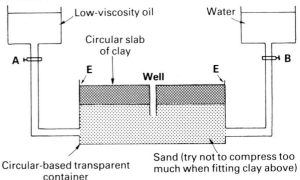

(h) Keep a project book of newspaper articles about off-shore gas and oil developments around Britain.

(i) Do a survey of building stones in your local town centre.

(j) Hand polish the surface of some rock specimens using progressively finer grades of silicon carbide powder.

(k) Visit civil engineering sites. Investigate the bed rock of the site. Draw maps and graphs to show sources/amounts of materials used in the construction.

(l) Make brick samples from clay baked in a pottery kiln.

(m) Investigate and measure the porosity of sediments (how much water they can hold). This can involve (i) finding the difference in mass between dry and saturated samples, (ii) measuring the amount of water that can be absorbed by a beaker full of sediment. Try different materials (vary the grain size, the amount of compaction, etc. and have some samples already cemented by slurries of Polyfilla). The way porosity works can be demonstrated by seeing how water fills the gaps in a container packed with marbles, ball bearings, etc. of various sizes.

Index

Important page references are printed in **bold**.